RETHINKING THE CONCEPT OF WASTE AND MASS CONSUMPTION

This book presents hard facts, drawn from extensive research, to highlight our unsustainable consumption of the Earth's resources and the limitations of the UK's current management of waste and recycling.

Setting out a bleak picture of a world in which we are literally consuming our planet, the book explores the psychological, economic and capitalist drivers behind this behaviour. Controversially, the book examines the drawbacks of the current approach adopted by many local authorities on the kerbside collection of recyclable materials, as well as the UK governments' strategic approaches to household recycling, including the lack of UK-wide infrastructures for packaging reuse, and for product repair and recycling. It challenges the whole concept of waste, leading to a proposed new strategy for the management of household waste, including a simplified household collection system, the introduction of an incineration tax and the banning of all household waste exports. The author proposes reconceptualising waste as unwanted but valuable material and argues that the responsibility for facilitating reuse, repair and recycling, rests with manufacturers who must start designing with the end in mind.

Given the current economic climate, and a dampening of the green agenda within UK politics, the book provides a much-needed call for critical discourse on how, and how much, we consume and sets out clear, practical solutions for change. The book will be of interest to manufacturers, retailers, consumers, local authorities, policy makers, students and professionals looking to reduce our impact on the environment.

Richard Waite studied and practised as a chartered engineer, before setting up one of the UK's first household recycling schemes in the late 1980s. Subsequently, as a management consultant, he advised many councils and the Government on household waste management and recycling, and in 1995 wrote the first book on household waste recycling. He was the specialist advisor to the House of Commons Environment Select Committee during their 1993/94 inquiry into recycling. He then became the Managing Director of several very successful UK businesses but maintained his interest in recycling. Drawing on his own experience setting up and running one of the UK's first companies recycling Household Waste, plus extensive research and analysis, he has written this book to draw attention to the crisis we face and to present a comprehensive blueprint for how we should in future manage our use of the Earth's limited resources.

Routledge Studies in Sustainability

Waste and Discards in the Asia Pacific Region
Social and Cultural Perspectives
Edited by Viktor Pál and Iris Borowy

Digital Innovations for a Circular Plastic Economy in Africa
Edited by Muyiwa Oyinlola and Oluwaseun Kolade

Critical Sustainability Sciences
Intercultural and Emancipatory Perspectives
Edited by Stephan Rist, Patrick Bottazzi and Johanna Jacobi

Transdisciplinary Research, Sustainability, and Social Transformation
Governance and Knowledge Co-Production
Tom Dedeurwaerdere

Digital Technologies for Sustainable Futures
Promises and Pitfalls
Edited by Chiara Certomà, Fabio Iapaolo, and Federico Martellozzo

Rethinking the Concept of Waste and Mass Consumption
Preserving Resources through Reuse, Repair and Recycling
Richard Waite

For more information on this series, please visit: www.routledge.com/Routledge-Studies-in-Sustainability/book-series/RSSTY

RETHINKING THE CONCEPT OF WASTE AND MASS CONSUMPTION

Preserving Resources through Reuse, Repair and Recycling

Richard Waite

Routledge
Taylor & Francis Group

LONDON AND NEW YORK

from Routledge

Designed cover image: MarcelC © iStock

First published 2024
by Routledge
4 Park Square, Milton Park, Abingdon, Oxon OX14 4RN

and by Routledge
605 Third Avenue, New York, NY 10158

Routledge is an imprint of the Taylor & Francis Group, an informa business

© 2024 Richard Waite

British Library Cataloguing-in-Publication Data
A catalogue record for this book is available from the British Library

ISBN: 978-1-032-82487-1 (hbk)
ISBN: 978-1-032-82486-4 (pbk)
ISBN: 978-1-003-50475-7 (ebk)

DOI: 10.4324/9781003504757

Typeset in Optima
by Newgen Publishing UK

CONTENTS

FIGURES AND TABLES

Figures

Tables

INTRODUCTION

Why I've written this book

When I started writing this book, I was thinking about how to improve our current approach to household "waste" management and how we need to manage our Household Waste differently.[1] But I quickly realised that this is the wrong way of looking at the problem (the problem being that we are consuming the natural resources of our planet at a relentless rate and then adding insult to injury by polluting the planet with "waste" when we no longer want things). By focusing on how to manage our Household Waste better, I was looking at the symptoms of the problem, rather than at the underlying cause, which is our culture of excessive consumerism. I was looking through the wrong end of the telescope.

Throughout this book I have put the word "waste" in inverted commas, because "waste" is really just stuff we no longer want, but which still has an inherent value. I have put it in inverted commas because these items are not "waste" at all, as I will explain in the next chapter.

I know quite a lot about managing Household Waste, but I didn't know so much about consumerism and what drives it. So, I have read and researched widely and have tried to pull together my own thoughts, with the ideas and views of some of the leading thinkers on this issue, people who are, in the main, enlightened economists. I say enlightened, because as you will see in Chapter 7, one of the main causes of our problem of excessive consumption is our economic system and specifically our obsession with and reliance on, economic growth.

So, the issue we need to address is actually that of consumerism and our rampant consumption of the Earth's limited resources. We have to find a new

DOI: 10.4324/9781003504757-1

approach, which I have come to think of as "sustainable consumerism" and no, this is not an oxymoron, consumerism can be sustainable as I intend to show.

Sustainable consumerism is different to our current pattern of consumerism in two ways, in that it requires:

- changes to the products and packaging that we buy and changes to how we buy; and
- changes in the way we treat what we no longer want (the so-called "waste").

But, when talking about consumerism, you might think the solution is all down to you and me, as the consumers; far from it. For sustainable consumerism to succeed, we need a significant group of players, those organisations that comprise the traditional supply chain[2] and beyond, to all play their part. These players are:

- producers, by whom I mean: the manufacturers of durable goods (think fridge, TV, clothing and shoes, mobile phones); producers of consumable products such as food and drink and other consumables; and the producers of packaging;
- retailers (large and small) who sell us products;
- you and me as consumers;
- local authorities who collect and manage the treatment of our Household Waste on our behalf;
- the waste management industry which actually deals with our Household Waste;
- the reprocessors of our Household Waste who turn unwanted materials, such as paper and glass, into secondary raw materials that can be used to make new products; and
- the four regional UK Governments that set the policies, regulations and rules by which all the other players must operate.

I'll start with a disclaimer. This book is not a rigorous scientific analysis of Household Waste management. Neither is it an environmentally exhaustive list of how we should manage our use of natural resources. I don't claim to have all the answers; there are many people better qualified than me to provide the missing answers. But where I don't know an answer, I hope at least to be asking the right and, at times, blunt questions.

I have tried to present a pragmatic set of proposals for a new approach, based on common sense and an understanding of how much (or rather how little) effort people will make when it comes to throwing stuff away. And I'll start by being brutally honest; we are all guilty of throwing stuff away, since doing anything else with it is often just too much trouble; it's too hard compared with chucking it, which is easy. But, we could all make a little more effort, indeed we have to, because how we all currently behave is unsustainable.

By adopting the proposals I will present, we can start to address the urgent need to reform how we all consume stuff in our lives, because we are currently wasting the Earth's limited resources. So, how should we be dealing with what we no longer want?

We're doomed

For those of you old enough to remember the television series "Dad's Army", the character Private Fraser had a favourite phrase when things looked bad; he would say "We're doomed Captain Mainwaring, we're doomed". And humankind is doomed, unless it radically changes its ways and quickly. We all know about the urgent and catastrophic impact of climate change and global warming and there are many other environmental and political challenges facing our species, from loss of biodiversity and energy shortages to war. But I want to focus on one issue that is not getting the same attention as these bigger environmental issues, but which nonetheless is having a major and detrimental impact on our world.

This issue is that as global consumer societies, and I'm thinking here of all developed and developing countries, not just the UK. We are consuming the Earth's limited natural resources at an unsustainable rate and trashing the planet by discarding things when we no longer want them. This consumerism is slowly but surely degrading and exhausting our planet and must change.

In this book I have focused on the UK, but the issues and my proposed solutions are applicable to any developed country that operates house-to-house Household Waste collection (e.g. Europe and North America).

You may well say, but the UK is only a very small part of the global economy and I am only one of more than 63 million people living in the UK[3] (and only one of nine billion globally),[4] so how can I make a difference? Whilst we may all feel small and insignificant in the face of the global challenges facing our planet, we are all important. How each of us behaves does make a difference, both in terms of our individual consumption and our ability to influence others. Greta Thunberg called her recent book *No One Is Too Small to Make a Difference*[5] and look at the impact she has had.

In his book *A Life on Our Planet*, Sir David Attenborough talks about the town of Pripyat and the nuclear disaster in April 1986, otherwise known as Chernobyl. He says:

> We live our comfortable lives in the shadow of a disaster of our own making. That disaster is being brought about by the very things that allow

us to live our comfortable lives. And it is quite natural to carry on in this way until there is a convincing reason not to do so and a very good plan for an alternative. That is why I have written this book.[6]

And this is exactly the reason I have written **this** book. I think there is a convincing reason to change and I will present what I think is a good plan for an alternative.

I'd like to quote another author, Brigit Strawbridge Howard, who in her book *Dancing with Bees* wrote: "And whilst many of us view this planet as a wondrous, magical place to be loved, cherished and cared for, others view it as nothing more than a 'resource', something to be managed, controlled and exploited – then discarded when it no longer serves its purpose".[7] I happen to share Brigit's view that the Earth is a wondrous, magical place to be loved, cherished and cared for and I challenge anyone not to feel this way if they take the time to truly experience the wonders of nature that exist all around us.

However, I also understand that many people do not share this view, they lead busy lives, they have other priorities and, sadly, too many of us do see the Earth as a bunch of infinite resources to be exploited for human convenience. But these resources are not infinite, we can already see where there are limits, plus throwing away "waste" is polluting and damaging our once pristine planet in so many, often irreversible, ways.

But, where I disagree with Brigit is that I think there is a third group of people who just don't think about our planet and what we are doing to it; they simply don't care. And sadly, I suspect this group is in the majority. You only need to look at the statistics on litter in the UK. According to an article published in the *Sunday Times* newspaper,[8] in 2015 the United Kingdom was the most littered country in the Western world. The most littered!

The 2017 Government Litter Strategy for England stated that three quarters of a billion pounds was spent each year by local government clearing up litter. It also stated that "81% of people were angry and frustrated by the amount of litter lying all over the country";[9] however, according to the Sunday Times article, 48% of people surveyed admitted to littering. This is perhaps a good example of people who don't think or just don't care about what happens to materials they no longer want.

But, as I will explain in a moment, I'm not just talking about how we manage our Household Waste better, it is more fundamental than that. What needs to change is how we consume stuff. Okay, that sounds like a huge and perhaps unachievable objective, but stay with me, I do think it could be done whilst still allowing us all to consume new products. As Baldrick used to say in the television series Blackadder, "I have a cunning plan".

I want to show committed environmentalists and people who care enough to recycle and also those people who don't really think or care about what we are doing to our planet, not only that we **have** to change how we consume, but **how** we should change.

The very first thing I want to say is that whatever we do to change things, we have to make the new approach simple or the majority of people just won't do what is required. We cannot make things too difficult to understand and we must not make things so complicated that people just won't bother. So my "good plan for an alternative", as Sir David calls it, is based on the KISS principle (Keep it Simple Stupid). And I'm not the only one saying this.

Miller and Aldridge in their book *Why Shrink Wrap a Cucumber?* say "The popularity and use of recycling depends to a great extent on user-friendliness, so for households co-mingled kerbside collections[10] are often preferable to local recycling centres".[11] I will explain in Chapter 4 how our Household Waste is currently collected and then in Chapter 10 I set out why I think we should move to co-mingled, kerbside collection of Dry Recyclables.

Miller and Aldridge go on to say:

> Some environmentalists and end-users of these materials argue that co-mingled recycling results in an inferior product because of cross-contamination, but this has to be balanced against the convenience to the consumer. In the UK it is not uncommon for local authorities that switch from sorted waste (that is, asking householders to separate their Household Waste into as many as five or six different boxes or bags) to co-mingled waste collections, to experience a 20% increase in recycling volumes.

This is why I think we have to apply KISS to recycling collection.

Now, I accept this isn't the first book to be written on the subject of recycling (in fact I wrote one in 1995). Much has been written recently about the problems we face with regard to Household Waste and plastics in particular. Most of these books explain the problems and offer advice on how we, as individual consumers, can allegedly change our patterns of consumption to help address these problems.

This is all good stuff, but as consumers we can't solve the problems on our own. Of course we must all play our part, but we need a national, co-ordinated plan (and action) involving all players involved in the production, consumption and post-use treatment of products and packaging. And the changes have to be made at scale. By this I mean the changes need to be national, not local or individual, otherwise we will just be tinkering at the edges of the problem. So, my first message is that if we are to change our current, unsustainable ways of consumption, the changes we need to make must be:

- simple for the consumer/householder to understand and act on (KISS);
- implemented at scale; and

- enacted by all the players involved in the production and management of consumer products and packaging, including when these become unwanted.

Can we bring about such change? Yes, I believe we can, indeed it has to be done, otherwise future generations will inherit a planet that has been stripped of key resources and is drowning in "waste".

So, the purpose of this book is not just to look at what you and I can do differently as consumers and householders, but to present a complete picture of what needs to be done and by whom (producers, retailers and Government, both local and central, consumers and the waste management industry), to change how we manage our limited resources.

Whilst we as consumers can change our buying and disposal habits, we need producers to enable us to do this by designing, making and packaging products in different and simpler ways and we need retailers to encourage producers to do this. Then we need the infrastructure to be created that will manage and process the mountains of recyclables, compostables and recoverables that we will create, instead of just throwing stuff away. It would be a big change, but I honestly believe it is doable and I've set out in detail how it can be done, including how I think it could be funded.

But most of all I want us to stop using words like "waste", rubbish, garbage, trash and refuse. Instead, I want us to talk about "unwanted materials", because this is precisely what "waste" is; it is materials that we once wanted, but now no longer want. They are simply materials in the wrong place, at the wrong time and often in the wrong form to be of immediate further use. So, I actually want us as a society to practise Household Waste reduction, reuse, repair, recycling and recovery.[12]

It's interesting that so many of the terms for more sustainable practices start with the letters "re". This is perhaps not surprising as "re" at the start of a word often means to do again, for example: restore, revive and refurbish. We cannot go on using the Earth's precious resources just once and then, when we have had enough of them, consign them to the great dustbin that is a landfill site or a fiery inferno that is an Energy from Waste (EfW) incinerator (see Chapter 5). We have to up our game in terms of Household Waste reduction, reuse, repair, recycling and recovery if we are to leave this planet in at least as good a condition as when we were born onto it.

As you read this book, you will notice that I refer to products and packaging as two different things. Just to be clear, by products I mean the things that we buy because we want them for a particular purpose. This might be a television, an item of clothing or a food item.

Clearly, when I'm talking about packaging, these are the materials wrapped around the products that we buy, to protect them, keep them safe and to allow the producer to tell us about the product, its features, how to use it and certain statutory information. The key difference between products and packaging is that packaging is inherently transient. Once we've got our newly purchased product home, we usually discard the packaging straight away as it is no longer needed. The differences between how we should treat products and packaging will become clearer as we progress, particularly when I start to talk about options such as reuse, repair and recycling.

Who am I to talk about consumerism and Household Waste?

You might ask, who am I to say all this? Well, I have a long-standing and some might say somewhat bizarre interest in Household Waste.

Maybe my interest started when I worked as what was then called a dustman, back in 1976, between school and university. That was when households had steel dustbins that we had to carry on our shoulders from the back of the house to the dust-cart and tip the contents in. None of today's wheeled bins, put out by the householder at the kerbside and mechanical lifts on the back of the lorry to empty the bin. Today's Household Waste is untouched by human hands; not then.

But, my interest really kicked in during the 1980s when I became increasingly concerned that everything we no longer wanted was just being thrown away. Household Waste recycling hadn't really started in any meaningful way. So, I gave up my job and set up one of the UK's first kerbside recycling collection schemes in partnership with Leeds City Council. The Council collected recyclable materials in special divided wheeled bins and lorries and I built a sorting plant (a simple Materials Recovery Facility or MRF)[13] to which they delivered the recyclables. One of the purposes of this partnership was to prove that kerbside collection of mixed (or "co-mingled") recyclables could work and to gather data on the make-up of Household Waste and the amounts that could be successfully collected and sorted.

My plan was to prove the concept and then get a big waste management company to roll out the approach. Unfortunately, my timing was all off, because as soon as I was ready to talk to the big players, the '80s recession had kicked in and then was not the time. So, after my one and only ever sleepless night, I closed down the operation, sold off all the sorting equipment and my car, took on personal debt and got a job. But the job I got was as a consultant advising councils and the Government on Household Waste management and recycling. I somehow found the time to write what was then the only book on Household Waste recycling,[14] advised many, many councils on recycling and was a Specialist Advisor to the House of Commons Environment Select Committee during their 1993/94 inquiry into recycling. I was also one of the three authors of the original Landfill Tax proposals.

So, I've got form in the field of Household Waste recycling and "waste" management. As a keen recycler myself, I think I know what you can and can't do today. But many people don't; they don't know what can be recycled, they don't know how best to present unwanted materials for recycling and many simply don't believe that stuff actually gets recycled (they think it all goes to landfill or incineration). But, what really frustrates me is that we could be doing so much more. No, that's wrong; we **have** to do so much more.

Spurred on by a real fear that we are desecrating our planet; for example as seen by the horrors of plastic waste pollution in our oceans (as so vividly depicted in the excellent BBC series Blue Planet II presented by Sir David Attenborough), I decided to try to bring about change by writing this book. We need to look very differently at how we manage Household Waste, starting by recognising that it is not "waste" at all, it's just stuff we no longer want.

I started writing about how we could better manage our Household Waste but quickly realised that I was writing about treating the symptoms rather than the cause of the problem (the cause being consumerism). Yes, we can improve the ways in which we manage our Household Waste, and I'll talk later about how I think we should do this, but much more importantly, we have to stop creating so much. And, as I've said, we have to change how we consume the Earth's resources.

I've had to add to my knowledge of Household Waste management and have therefore read widely and researched the subject of consumerism, which has led me into the complexities of how our economy operates and the fundamentals of capitalism. I have tried to pull together the ideas and views of some of the leading thinkers on this issue, who are, in the main, enlightened economists. I say enlightened because, as you will see in Chapter 7, one of the main causes of our problem of over-consumption is our economic system and specifically our obsession with and reliance on consumptive, economic growth.

I'll give you just one example of this. When our recent, short-lived UK Prime Minister Liz Truss came into office, her whole approach to solving our country's dire economic position was summarised in four words "economic growth, growth, growth". The way to solve all our problems was to "get the economy growing again", then all would be well. Oh, and whilst we're at it, let's row back on many of our environmental protection plans to make economic growth easier and faster. I despaired. As you will see in Chapter 7, this thinking is so flawed as to be unbelievable, simply because our current definition of economic growth relies totally on growth in the consumption of natural resources, which are finite. Our planet is finite. As Professor Tim Jackson said in his seminal book *Prosperity without Growth*[15] "we are running out of planet".

Now, I make no claims to be an expert in these fields, I am certainly not an economist. But it seems to me that you don't actually need to be an expert to understand the problems we are facing. But, we do need experts to figure out how we can fix the problems and, as you will see, the solution is not simple and its achievement will not be easy. There are huge political issues to address, plus societal changes and behavioural changes, not to mention revisiting how we measure and hence drive economic growth. Heavy stuff.

But, whilst we need to task the experts with addressing all this geopolitical, economic thinking, there are practical steps we can and must take now, to at least ameliorate some of the damage that we are currently inflicting on our beloved planet.

So, for the next six chapters I am going to look at how we consume the Earth's resources today and what the problems are with our current approach. Then in Chapter 7 I will present my understanding of the economic and social drivers of our current unsustainable consumption. Then I will focus on the specific issue of packaging in Chapter 8 and on plastics in particular in Chapter 9, before proposing solutions in Chapter 10, along with estimates of what we could achieve if we were to change in Chapter 11. If you're still with me by Chapter 12, I've set out a high-level plan for what I think we need to do differently now, not tomorrow, not next year or by 2030, but now. The issues we are facing are that serious.

Notes

1 I have defined what I mean by Household Waste in Chapter 2, but for now, please think of it as the contents of your wheeled bin or black refuse sack and anything you take from your household to your local Household Waste Recycling Centre.
2 The traditional supply chain is all of the players and their actions that take raw materials, turn them into products, package them and sell them to retailers who sell them to us as consumers. It does not include those players and their actions once a consumer discards no-longer-wanted products or packaging.
3 Office for National Statistics, "2011 Census: Population Estimates for the United Kingdom, March 2011", ons.gov.uk, accessed 29 September 2021.
4 U.S. and World Population Clock, www.census.gov/popclock/world, accessed 29 September 2021.
5 Greta Thunberg, *No One Is Too Small to Make a Difference*, Penguin, 2019.
6 David Attenborough, *A Life on Our Planet*, Witness Books, 2020, p. 7.
7 Brigit Strawbridge Howard, *Dancing with Bees: a journey back to nature*, , Chelsea Green Publishing, 2019, p. 209.
8 David Collins, "Britain's become a rubbish dump. It's up to us to pick up the pieces", the *Sunday Times*, 28 March 2021.
9 HM Government, "Litter Strategy for England April 2017", Summary, p. 9.
10 Co-mingled collection is when all Dry Recyclables (the paper, plastics, cans and glass bottles and jars, etc.) are all mixed up together in a single container, normally a wheeled bin. Kerbside collection is when the householder is required to set out their Household Waste for collection, at the curtilage of their property, usually the pavement outside their property or the road at the end of their driveway.
11 Stephen Aldridge & Lauren Miller, *Why Shrink Wrap a Cucumber? The Complete Guide to Environmental Packaging*, Laurence King Publishing, 2012, p. 50.
12 This is the so-called EU Waste Hierarchy, as I will explain in Chapter 3. However, at the end of Chapter 3, I propose a replacement hierarchy, which I have called the Materials Management Hierarchy or the Three Rs.

13 Pronounced "Murf", this is an industrial facility that sorts mixed recyclable materials into single material streams and then sends these on for reprocessing. Some sorting is carried out by hand, but the vast majority is automated. See Chapter 5 for more on this.
14 Richard Waite, *Household Waste Recycling*, Earthscan, 1995.
15 Tim Jackson, *Prosperity without Growth*, Routledge, 2017, p. 17.

1

CONSUMING THE EARTH'S RESOURCES

So, what's the problem?

As I've already said, we're all aware (or should be) of the cataclysmic changes and risks facing our planet: climate change, over-population and the massive loss of biodiversity.[1] But, I've written this book to draw attention to another very damaging impact that humans are having on our planet, which is that we are consuming the Earth's limited natural resources at an unsustainable rate and trashing the planet by simply discarding things when we no longer want them. And this is all driven by our seemingly insatiable consumerism, which is slowly but surely destroying our planet. So, we have to change.

I've written this book to:

- draw attention to this urgent and potentially catastrophic issue;
- explain what happens now and why this isn't good enough; and
- show what would realistically be achievable if all the players involved, including you and me, make changes to facilitate sustainable consumerism.

How consumerism actually works

Figure 1.1 illustrates how we consumers devour the Earth's natural resources. Three basic activities start this process. Companies mine, extract or harvest the raw materials they want. Mining in particular is not only a very environmentally damaging process, but also a very wasteful one, as the amount of the material that miners want is often only present in very small quantities in a much larger amount of the naturally occurring ore that is mined. Witness the slag heaps in Wales and Yorkshire from coal mining or the

DOI: 10.4324/9781003504757-2

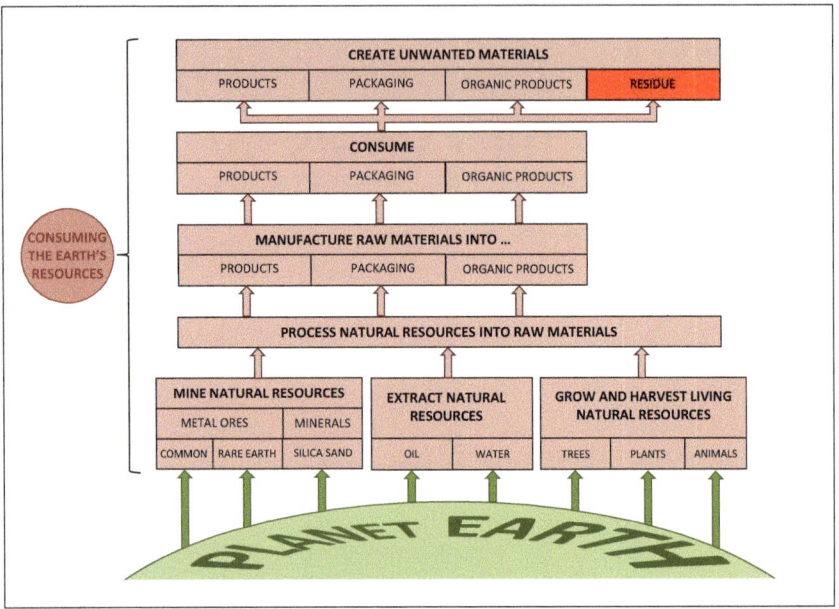

FIGURE 1.1 Consuming the Earth's resources.

white mountains in Cornwall left from china clay extraction. Extracting oil is no better and harvesting living natural resources is a moral minefield.

These natural resources are then processed into raw materials ready to be turned into products, packaging or food.

Once this has been done, we the consumers buy what we want and either literally consume them (as food or drink) or use them until we no longer want them and what we no longer want we dispose of; out of sight, out of mind. And until relatively recently, this simply meant burying it in a convenient hole in the ground, i.e. landfill.

Consumerism takes the Earth's natural resources and turns them ultimately into what most people, from the Government and local authorities to us as householders, call "waste" or rubbish.

But it's not "waste"

What we are actually talking about are materials or products that we no longer want or need, but which still have an intrinsic value; they are not "waste". We need a different word to describe this stuff, because "waste" implies they have no value and need to be thrown away, got rid of or disposed of and we pay someone else to do this for us (our local council).

We practise out of sight, out of mind. Once we've put our "waste" in "the bin" or taken it to what is euphemistically now called a Household Waste Recycling Centre (HWRC), but what used to be called "the tip" or "the dump", we forget about it and think it's no longer our problem. But it is still our problem, because what happens to these unwanted materials is actually our responsibility. We consumed them in the first place and we should make sure that they are treated in an environmentally acceptable way.

But, if it's not "waste", what is the word to describe what we no longer want? The first bit is easy, they're materials. They are unwanted or no longer needed, but they could be reused, repaired, recycled or recovered in some way. They are no longer raw materials as they have been converted into something useful. So, what do we call them instead of "waste"? I think we should call them what they are: "unwanted materials". They are clearly unwanted otherwise they wouldn't have been thrown out and they definitely remain materials. So instead of a "waste" or rubbish bin, we have an "unwanted materials bin". Not catchy I grant you, but at least it's honest.

But, I actually want to get rid of the whole concept of Household Waste, starting with the word "waste". Waste is a concept created by humans; it doesn't exist in nature. In the natural world, the output of any process becomes the input to another process. So, an animal eats food and produces bodily waste products (an output), which fall to the ground and are broken down by microbes in the soil, producing nutrients (inputs) that are taken up by plants to help them grow. These plants are eaten by animals or humans and the cycle continues. This is a circular pattern.

What humans have created is a linear pattern where natural resources (wood, metal ores, oil, etc.) are refined and then turned into products. You and I, as consumers, buy these products and when we've had enough of them we discard them and expect someone else to get rid of them for us. "Getting rid of" can mean recycling, but more usually means landfilling or, increasingly, incineration. So, it's a one-way ticket from extraction to disposal; it's linear not circular. And, because we want more new products, there has to be more raw material extraction rather than reusing the materials that have been discarded.

But it's not just you and I as consumers who create unwanted materials, it happens all the way through the processes shown in Figure 1.1. From the slagheaps to the unwanted by-products (often toxic chemicals) of refining natural resources into raw materials, to the unwanted rejects and offcuts created during manufacturing, we create stuff we don't want, all along the way.

As a result, the materials that end up in our finished products represent a surprisingly small amount of what we originally took from the earth. Now this might be okay if it weren't for the fact that having taken what we want, we then return what we don't want to the earth and this can often be in unnatural or even toxic forms. And it's also very much not okay when the natural resources are finite and irreplaceable, as many are.

Some people say that we humans should mimic nature, so that we eliminate the concept of "waste" by using any unwanted material as the input to a new product. Whilst this sounds sensible and attractive, it is rather simplistic. Yes, we can take unwanted paper and cardboard or metal cans or glass bottles and turn them back into useful products through recycling. But we can only do this because these are single materials.

What humans are very good at is combining materials to produce more complex products that comprise a mixture of materials. At a simple level, this might be the plastic window in cardboard food packaging or multi-layered packaging such as drinks cartons that combine layers of paper, plastic and even aluminium. At the top end of this scale we have electronic products that combine multiple types of plastic, metals and other materials such as glass, or think about a car!

Unless we can break down these complex products into individual materials, we can't recycle them.

And there are other products that are just too difficult and therefore too expensive to recycle, for example used nappies, filled dog-waste bags, cigarette butts or the contents of your vacuum cleaner.

So, a key question is: do we have to accept that some of our unwanted materials cannot or will not be treated in an environmentally acceptable way, i.e. in a circular manner? I'll come back to this later.

Being a throw-away society is not sustainable

We are a convenience-driven, consumer society. We take, make, consume and then waste things. We discard some things when they are no longer of value to us, because:

- they've served their purpose, for example packaging;
- they break or wear out;
- a better one comes along; or
- we're just bored with them.

This is today's consumerism and it's not sustainable.

But, there is a part of consumer convenience that is even worse than discarding stuff we no longer want and this is about making and consuming things that are planned to be disposable; we make and buy things with the intention of throwing them away. Disposable is seen to be a good thing, an acceptable thing. Examples are:

- disposable cutlery, plates, glasses and of course take-away coffee cups, that don't have to be washed up;

- plastic film bin liners to contain our "waste" before putting it into another "waste" container;
- tissues instead of handkerchiefs to save washing (but you could at least argue these are more hygienic);
- single-use plastic drinking water bottles;
- take-away food cardboard packaging and food containers such as pizza boxes and polystyrene clam-shells;
- disposable nappies, again to save washing; and …
- the list goes on.

Part of being a convenience-driven consumer society is our acceptance that disposable is normal and acceptable; but this cannot continue. Whatever the reason for the disposal of stuff, there has to be a better way and we have to change.

Holding up a mirror

We have become a consumer society and that means we consume resources. The *Oxford English Dictionary* definition of "to consume" is: "to eat or drink; completely destroy, reduce to nothing or tiny particles …". It's not a very flattering description of a consumer society, but it's accurate (and remember the "tiny particles" bit when I come on to talk about plastics).

In recent years we have become a society that wants things instantly and conveniently delivered. This started with the creation of supermarkets and then came online shopping, based on the convenience and immediacy of the internet (we don't even have to leave home to shop), which was exacerbated during the Covid-19 lockdowns. I recently came across an extreme example of this when I read about a man in Australia who has a cup of coffee delivered to his house by drone every day.[2] My response to such ultra-consumerism is to say: **just because you can, doesn't mean you should**. Please think about the consequences of your actions before making your purchasing decisions.

A big part of our consumer society is that we want things pre-packaged for convenience, for example food and drink, particularly preprepared food, and when we've finished with things we throw them away (first the packaging, then ultimately the bits of the food product we don't want).

A large part of what drives this is the on-the-go society in which we now live. How often do you see someone carrying a disposable cup of coffee as they walk along a street or a single-use plastic bottle of water? People used to make their own lunch to take with them to work; not anymore. Virtually everyone in a city buys pre-packed food and drinks for their midday meal. All of these products are designed for immediate consumption. So, the packaging is quickly discarded and the nature of the packaging makes it very hard to recycle; and anyway it's rarely collected for recycling, it all goes into mixed litter bins, the contents of which are destined for incineration or landfill.

And to most consumers, price is everything, quality is secondary; cheap is good enough, because that way we can buy more stuff.

As I say, it's not a very pretty picture, nor one we should be proud of. Try explaining to a young child how our consumer society works and see if they are impressed.

Simpler times

It used to be that we bought less. Okay, there was less choice and less to buy, but is today's plethora of choice necessarily better? We made some things ourselves, like clothes, but home sewing machines have nowadays largely gone out of fashion.

We saved up for things and waited to buy them and it was accepted that you paid more for quality products that lasted. And we repaired things when they wore out or broke and products were manufactured in a way so that they could be repaired.

Yes, I know it's easy to say that this is looking at the past through rose-tinted glasses, but stay with me awhile, maybe there are things we should learn from the past.

What changed

First, the price of goods became dominant. And this was largely, if not exclusively, driven by producers. Manufacturers changed to making products as cheaply as possible, in order to out-compete their rivals on price, and as consumers we lapped this up.

But, reducing the price of a manufactured item has often been achieved by reducing the quality of the product. Things don't last as long and are nowadays made in a way that means they are no longer repairable. This is again to the producer's advantage, because we consumers have to buy a new product to replace the lower quality one that has failed. This is the essence of what is called "built-in obsolescence", which is the practice of deliberately designing and manufacturing products to fail after a certain period of time, in order to force the consumer to buy another one.

Built-in or planned obsolescence has its roots in the Great Depression of the 1930s and was built upon in the 1950s in the aftermath of the Second World War. It was an approach adopted by producers to increase consumer demand, either by making products that had a finite life in terms of their appeal (achieved by constantly bringing out new designs, features and attributes, making people want the latest and shiniest models) or by making the products physically fail after a limited amount of time, so that the consumer had to buy a replacement.

I remember as a boy hearing how the bodywork of Japanese cars was known to be of inferior quality to other countries' cars and that they rusted quickly. But because of the poorer bodywork, such cars were cheaper to buy and so were popular (until they rusted). I should say that this is no longer true, the rusting that is, not the popularity.

And it's happening now with one of the most prevalent products on the planet – mobile phones. Most people are aware of the rumours that mobile phone suppliers deliberately slow down older phones during software upgrades as they pass the magical two years of age (to coincide with typical contract renewal periods). This creates dissatisfaction in the consumer with the phone's performance, which coupled with the constant launching of newer or better features (first it was storage capacity and battery life, now it's camera resolution) creates a desire in the consumer to want a new phone. The world's leading mobile phone provider has admitted the practice and has been fined in the US as a result.[3]

But going back to cars, nowadays, car manufacturers produce cars with much improved bodywork and long-lasting paint protection, so bodywork is no longer a source of built-in obsolescence (one in five cars on the road is now at least 13 years old, twice the proportion of a decade ago).[4] So, how do car manufacturers build in potential failures to encourage you to replace your car every few years (apart from fashion changes)? The answer is in the increasingly complex and prevalent electronics built into cars. When a simple electronic component fails, try finding someone who will diagnose the problem and replace the simple resistor, capacitor or microchip that has failed. All you will find is a repairer who says the only option is to replace the whole circuit board, because "it's not worth the cost of repairing the fault". And it's even worse because the circuit boards are unique to the car manufacturer, so you can't even shop around for standard parts, you have to buy from the specific manufacturer's supplier.

This is an example of the second major change that has occurred in consumer behaviour, in that people now prefer to throw away a product when it is worn out or broken and buy a new shiny one. The idea of repair has dwindled almost to non-existence. So, not only are producers making products that are difficult or impossible to repair, but consumers have grown to prefer buying a new item instead of repairing. And in large part this is because fashion has become all-consuming. Many consumers want the latest design, the newest look or the most up-to-date technology. Buying the newest/latest product has in many instances become the be-all and end-all, just think of mobile phones and fast fashion in clothing.

Two things have been the major drivers of this change in consumer behaviour: the availability of cheaper goods and the introduction of inexpensive, easily available credit, which encourages us to buy more as we no longer have to save up for things.

A third, huge, and I mean huge, contributor to this change has been the advent of plastics (see Chapter 9). Plastics have changed our consumer behaviour, because plastic goods are cheap to make, cheap to buy and are often designed to be disposable.

The impact of the majority of the world turning into a throw-away society has been two-fold:

- we are now consuming the Earth's resources at an increasingly unsustainable rate; and
- we are running out of ways in which we can dispose of the mountains of no-longer wanted products and packaging that are being discarded and we are polluting the planet with our unwanted "wastes".

Consuming the Earth's resources at an unsustainable rate

None of our planet's physical resources is infinite. Oil (for plastics), metals, sand (for glass), trees for paper and cardboard, and plants for textiles are all finite (you can only grow so many trees and crops, because the amount of productive land is finite). Our global population is increasing and so therefore is consumer demand. But, we cannot continue to just plunder the planet's finite resources. We have to move to what is now described as a "circular economy", or as near circular as possible (see Chapter 6 for a discussion of the circular economy).

To put our consumption of natural resources in perspective, Ed Conway in his book *Material World* stated that "In 2019 … we mined, dug and blasted more materials from the earth's surface than the sum total of everything we extracted from the dawn of humanity all the way through to 1950 … In a single *year* we extracted more resources than humankind did in the vast majority of its history … Nor was 2019 a one-off".[5]

Boy, is our consumption of raw materials increasing, and rapidly.

Consumption of non-renewable resources

This is particularly true for non-renewable resources such as oil and metal ores (from iron ore for steelmaking to bauxite for the production of aluminium to rare earth metals used in electronics manufacturing). Obviously, some finite resources are more abundant than others, but this does not mean we should squander them.

You may say I'm being alarmist, but I don't think I am. During a recent webinar I attended presented by the Ellen MacArthur Foundation,[6] a slide was put up which showed all the elements of the Periodic Table and highlighted those that are forecast to be depleted in the next five to 50 years. Based on known reserves and current rates of extraction, included in this list of 22 elements were: zinc, silver, cadmium, tin, tungsten, platinum, gold, bismuth and uranium. These are just the ones I immediately recognised. Am I being alarmist? Here's a forecast that we'll run out of new silver, gold and tin within 50 years! All of these metals and many more can be easily recycled once they have been separated out from the products that contain them, thus reducing the need for new mining. So, why isn't more being done to recycle these precious metals? Why aren't more people shouting about this?

Drilling down a little further, let's take a quick look at one of the least abundant but increasingly critical group of metal ores: rare earth metals.

Rare earth metals

There are about 30 rare earth metals with exotic names such as germanium, tungsten, antinomy, niobium, beryllium and gallium. They share the following traits:

- they are found in nature associated with abundant "ordinary metals", like iron and copper, but in tiny proportions;
- production of these metals is on a minute scale compared with more common metals, so this is reflected in their very high prices; and
- they possess exceptional properties demanded by today's green technologies, from wind turbines to electric car motors and batteries to solar panels.[7]

All rare earth metals share two other common characteristics:

- there is a limited supply as they exist in very small quantities where found; and
- the extraction and processing of these metals is hugely environmentally damaging.

It is more than ironic that our so-called "clean" technologies, such as wind and solar power and "clean" transport in the form of electric cars, rely completely on rare earth metals that are so environmentally destructive to extract and process. When we in the UK and other Western countries pat ourselves on the back for being "green and clean", we ignore the environmental destruction that occurs overseas to allow us to do this. We have effectively exported the dirty business that allows us to bask in the glory of a green future.

By 2040 the world is expected to need four times as many critical minerals for clean technology as it does today.[8]

Given the scarcity and high financial value of rare earth metals, is it not imperative that these metals are recycled efficiently, once the products that contain them reach the end of their useful lives? But today, this is not happening.

Consumption of renewable resources

If we accept that some natural resources are limited in supply, it's tempting to think we should replace these finite resources with renewable resources. So, for example, can we replace some plastic products with the equivalent made from wood? Or could we replace oil-based textiles, so-called man-made fibres, with plant-based alternatives like cotton, hemp or bamboo?

Again, these renewable resources are not infinite either. To grow crops and trees that can be turned into useful materials, you need land and you need water. Both of these resources are finite, and if you grow trees or plants for product manufacture, you can't use the land they need to grow food, which will become an increasing priority as the world's population continues to increase.

So you see, simply replacing finite materials with renewable materials isn't as easy as it sounds.

Using the "waste" outputs from one process as the inputs to a different process

This is a much-quoted aspiration and often references what happens in nature. It sounds ideal, but is it possible in our modern, manufacturing world?

It is sometimes possible to use "waste" products from one process as inputs to another. A simple example is that of a toothbrush handle being made from the offcuts previously discarded from the manufacture of bamboo products. So, whilst it might appear that a renewable resource is being grown to replace a finite resource and the producer may cite this as an environmental benefit of their product, the reality is more prosaic. Yes, the bamboo is more environmentally acceptable than a plastic handle, but the real benefit is that the producer is taking an unwanted output from another process and productively using it as an input to make a new product.

This is just one simple example, but it relies on the different producers co-operating to transfer the "waste" from one process to another. How often is this going to happen?

Why "waste" disposal is problematic

Landfill

The second negative impact of becoming a throw-away society, is all to do with throwing things away when we no longer want them.

Historically, the preferred method of disposing of unwanted materials was, and to a large extend still is, to bury them in a hole in the ground. Mining, quarrying and mineral extraction conveniently created holes that could be filled up with "waste".

Tipping "waste" into holes in the ground gradually became known by the euphemistic term of "landfilling". But two things are now making landfilling problematic.

First, we're running out of suitable holes. Too many have already been filled and new ones are not being created fast enough to accommodate all the "waste" that is being generated.

Second, I say "suitable" holes, since you can't just fill any hole with "waste" today as people did in the past. The nature of the "waste" has changed, becoming much more polluting than in the past and our attitudes to pollution control have tightened and rightly so. Consequently, the environmental standards applied to landfill today to protect, in particular, ground water supplies, are such that the cost of landfilling has risen considerably.

It used to be that once a landfill site was full, it was capped off with earth and abandoned. Depending on the future demand for land, some capped landfill sites were built on and in some cases this has given rise to problems (see Chapter 5 – How our Household Waste is treated – Landfill).

With the advent of the Landfill Tax in 1996[9] landfilling has now become an expensive, as well as wasteful, treatment option.

Energy from Waste incineration

The increasing cost of landfill is one of the reasons that has led to a huge increase in Energy from Waste (EfW) incineration. Instead of burying our Household Waste, many Waste Disposal Authorities (see Chapter 3 under Local Authorities) have decided to burn it instead, in purpose-built EfW incinerators. The term EfW derives from the fact that heat generated from burning "waste" is converted into electricity, i.e. energy. For more detail on this, please see Chapter 5.

Aside from costing less than landfill (because of the impact of the Landfill Tax), one of the arguments for EfW incineration is that at least some of the energy inherent in the Household Waste is recovered, so that has to be a good thing, doesn't it?

Maybe, maybe not. Arguments rage about how environmentally acceptable EfW incinerators are. There are concerns about the toxicity of the gaseous emissions that come out of their huge chimneys, particularly dioxins.

Incineration doesn't of course make the input "waste" completely disappear. EfW incinerators produce ash, representing 15–25% by weight of the input "waste" processed[10] and whilst some of this ash is inert and can be used for low grade construction materials, some is toxic and has to be disposed of in tightly controlled landfill sites.

But aside from the environmental performance of EfW incinerators, there is the overriding issue that burning Household Waste precludes any recycling of the materials in that "waste", it's the end of the road and it is far from circular.

However, there is a far more insidious problem with EfW incineration. Because EfW incinerators are so expensive to build, the waste management companies that own these facilities need long-term contracts that guarantee the volume and, in some cases, the minimum calorific value of the "waste" inputs. This has led to many Waste Disposal Authorities tying themselves into long-term Household Waste contracts that require the Waste Disposal Authorities to deliver a guaranteed amount of Household Waste year-on-year for up to 25 years.

This means that satisfying these contractual commitments for EfW incineration can be in direct conflict with increasing recycling. Whilst we could and I think can recycle much more of our Household Waste, this will increasingly be limited, not by collection or reprocessing issues, but by the contractual commitments to feed these EfW incinerator behemoths.

We cannot allow EfW incineration to become the dominant method of treating our Household Waste in the same way that landfill once was. The solution would appear to be simple: first, a Government moratorium on building any new EfW incinerators and second, the introduction of an Incineration Tax, along the lines of the Landfill Tax. But more on this later.

Polluting the planet through recycling

But it's not just landfilling and EfW incineration that have a harmful effect on the planet, even recycling can have a damaging environmental impact, for two key reasons:

- the residue that is generated by recycling certain materials has to be dealt with; and
- not all recycling reprocessing is actually carried out in this country; unbelievably some of our Household Waste is exported for reprocessing.

Recycling residues

The recycling of some materials generates by-products that cannot be recycled and therefore require disposal. For example:

- paper and cardboard recycling produces unwanted printing inks that are removed from the "waste" paper and cardboard inputs, plus contaminants such as plastic windows and the broken paper fibre residues (see Chapter 5 and the discussion on paper recycling in Annex 1); and

- the washing of plastics and the soup of product residues inside plastic containers has to be treated in some way.

So, recycling isn't necessarily the silver bullet to solving our Household Waste problem that many people think it is. Recycling produces CO_2 and residues that have to be disposed of, either through EfW incineration or landfilling or both. But we need to keep things in perspective. Recycling is far better than virgin raw material extraction and processing and produces far less CO_2 emissions than EfW incineration or landfill. Recycling materials such as metals and glass also requires much less energy than virgin material production.

The undeniable truth is that the only environmentally acceptable way to reduce the impact of Household Waste on our planet is to reduce how much is generated by consuming less, by product and packaging reuse and by product repair. More on this in Chapters 3 and 10.

Exporting UK Household Waste has to stop

Some Waste Collection Authorities make their own arrangements for the reprocessing of some of the materials they collect through recycling schemes. In the past, this has included letting contracts to companies that export the materials to countries with lower labour costs and weaker or non-existent environmental standards, making the export of some materials for reprocessing a cheaper option than carrying out the reprocessing in the UK. This is true in the case of paper and is particularly true with regard to plastics.

This practice is thankfully now less common (but still happens) as most of the countries that did want our "waste" materials, no longer want them. A very good example is China, which at one time was so desperate for raw materials to fuel its burgeoning economy, that it was willing to buy our "waste" materials and reprocess them to feed Chinese manufacturing. But this stopped in 2018, when the world economic recession caused Chinese exports to fall and the demand for recyclable "waste" to all but collapse. In 2018 China banned the importing of plastic "waste".[11] Some other countries, such as Malasia and Turkey, continue to take some UK plastics and paper for reprocessing, but these markets may also be drying up. I personally think this is no bad thing. I actually think the Government should ban the export of unwanted materials to other countries. From an environmental perspective this practice is very bad as the receiving countries have much lower environmental standards than the UK, with much of the exported "waste" either being dumped or burned in an uncontrolled manner, as the small-scale reprocessing industries are overwhelmed.

But, as importantly, there is a moral issue. Surely the UK should clear up its own mess, rather than export the problem to other, usually less developed, countries which suffer as a result?

And if, when you see images of the obscene pollution of our oceans by plastic waste, you think, "well, at least it doesn't come from the UK", then think again. Some of it does come from the UK via this export trade and the calamitous, so-called, reprocessing facilities in the receiving countries which allow plastic "waste" to spill into rivers and then into the oceans.

This practice of exporting unwanted materials for reprocessing overseas, usually in developing countries, has led to the UK contributing to one of the world's key environmental pollution scandals, that of pollution of our oceans with plastics.

So we do contribute directly to this major problem and whilst market forces have reduced the practice of "waste" exports significantly, the only responsible way this practice can be stopped permanently is for our Governments to ban it and I urge them to do so.

So, the problem is …

Just to remind you where I started this chapter. The problem we are facing is that we are a consumer society, with a seemingly insatiable appetite for products (I discuss this more in Chapter 7). The production and packaging of these products is slowly, but inexorably, stripping our planet of finite and irreplaceable resources. But to add insult to injury, we then pollute our planet by discarding far too many of these products and packaging in environmentally damaging and unsustainable ways.

But before looking at how we need to change and what we might achieve, I want to look at what happens to our unwanted materials today.

Notes

1 *The Times Online*, "Humans wipe out 70% of animals in 50 years" News, 13 October 2022.
2 Bernard Lagan, *The Times Online*, "Bird attacks on drones force Google to suspend home deliveries", 22 September 2021.
3 BBC, news website, "Apple settles iPhone slowdown case for $500m", 2 March 2020, accessed 19 May 2021.
4 *The Times*, Environment online newsletter 28 February 2023.
5 Ed Conway, *Material World*, WH Allen, 2023, p. 15.
6 Ellen MacArthur Foundation, "Understanding the opportunity: What does the circular economy mean for supply chains?", webinar, 7 February 2024.
7 Guillaume Pitron, "The Rare Metals War", Scribe Publications, 2020, pp. 14–20.
8 HM Government, "Resilience for the Future: The United Kingdom's Critical Minerals Strategy", 2022, p. 4.
9 House of Commons Library, "Landfill tax: introduction & early history", 2009.

10 US Environmental Protection Agency, "Energy Recovery from the Combustion of Municipal Solid Waste (MSW)", www.epa.gov, 9 February 2023, accessed 13 March 2023.

11 Ian Tiseo, "Plastic waste trade in the United Kingdom – statistics & facts", statista. com website, 22 November 2021.

2

WHAT WE THROW AWAY

What is meant by Household Waste?

When talking about "waste" associated with consumers, there are two different types of "waste"; these are called "Pre-Consumer Waste" and "Post-Consumer Waste".

These two very different types of "waste" comprise materials that could be reused, recycled or recovered in some way, or that have to be disposed of.

Pre-Consumer Waste

Pre-Consumer Waste is "waste" that is generated by the industrial processes used to make the products or packaging that we buy as consumers. This is manufacturing "waste" and is typically very clean, of consistent quality and comprises only a single material. Examples are: offcuts from making aluminium cans; cardboard boxes or clothing; trimmings from prepreparing vegetables; or wood or textile offcuts from making furniture. Such "waste" was in the past often sent for disposal, but increasingly producers are recycling their internal "waste" to save money by turning unwanted trimmings or offcuts back into useable raw materials. To send such materials for disposal is not only poor financial practice by the producer, from an environmental perspective it is criminal. These are good quality, clean and easily collectable materials that just happen to be in the wrong form; so recycle or recover them!

There is Pre-Consumer Waste that is generated from manufacturing, but there is also Pre-Consumer Waste that arises in the supply chain that takes manufactured goods from a factory to a shop where it is sold. This is essentially bulk packaging, designed to protect the products during transit and to make

DOI: 10.4324/9781003504757-3

them easy to handle. Examples here are pallets and large cardboard boxes containing multiple products and a distressingly large amount of plastic film that is used to wrap products to make them easy to handle. We as consumers never see this Pre-Consumer Waste packaging, but there are thousands of tonnes of it generated every year and whilst a proportion is recycled, a large amount is sent for disposal, for example plastic pallet wrap.

Post-Consumer Waste

But the focus of this book is Post-Consumer Waste, the "waste" that you and I generate, once we have bought a product. Post-Consumer Waste and Household Waste are the same thing so from now on I'll talk about Household Waste and by Household Waste I mean two things:

- the contents of our dustbins or, more correctly these days, wheeled bins; and
- materials taken by householders to:
  - their local Household Waste Recycling Centre (HWRC); and
  - stand-alone recycling banks.

From now on I'm going to call this "Household Waste", but the Government has a slightly different definition. The Department for Environment, Food and Rural Affairs (Defra) calls the above "Waste from Households"[1] and uses the term "Household Waste" to mean something wider, i.e. all "waste" that derives from the public to include Waste from Households, "waste" from street bins, street sweepings and "waste" from parks. But I'm going to use Household Waste to mean "waste" deriving directly from households either in our bins or what we take personally to the local HWRC (see Chapter 3) or recycling bring bank, as it's an easier term to use than "Waste from Households".

In this book I'm not talking about industrial, commercial, building, agricultural or other "waste", but about the stuff we consumers no longer want. But this stuff, whether it is products or packaging still has a value; it is not "waste" as I'll explain below.

Today's Household Waste comprises many things including:

- discarded packaging;
- unwanted food "waste";
- garden "waste" such as dead leaves in the autumn, plant and tree prunings and unwanted plants including weeds and soil;
- consumer goods that are damaged, worn out or simply no longer wanted, such as books, DVDs, toys, crockery, kitchen appliances, tools;

- damaged, worn out or simply no longer wanted clothing and shoes and accessories like handbags;
- so-called Waste Electrical and Electronic Equipment (WEEE), e.g. mobile phones, computers, kitchen appliances, radios and TVs, electronic toys and gaming devices;
- discarded furniture and other home furnishings, from furniture to carpets and curtains, beds, lamps and cushions;
- hazardous "waste" such as unused and unwanted paints and solvents, exhausted batteries, surplus cleaning products, the contents of your vacuum cleaner bag;
- wood that has either been replaced with something newer and smarter or is left over from DIY projects;
- unwanted soil, rubble from DIY projects, plasterboard from building works; and
- more …

And it's a mess. Unlike the nice, clean single material that comprises Pre-Consumer Waste, Household Waste is everything we no longer want, all mixed together. And when one of the things we discard is food "waste", mixing this in with everything else creates an unpleasant and very difficult to deal with soup of "waste".

What we actually throw away

Our society produces many types of "waste". Table 2.1 shows a summary of the main categories of "waste" that have to be dealt with (the latest data available are for 2016).

As you can see from Table 2.1, Household Waste is a relatively small percentage (12%)[2] of the nearly a quarter of a billion tonnes of "waste" dealt

TABLE 2.1 "Waste" categories and UK arisings for 2016

"Waste" category	Million tonnes	Percentage
Construction, demolition and excavation (including dredging)	136.2	62%
Commercial and industrial	39.8	18%
Household Waste (what Defra calls "Waste from Households")	27.3	12%
Other	17.7	8%
TOTAL	**221.0**	**100%**

Note: Defra, "UK Statistics on Waste", 19 March 2020, Table 7, p. 11.

with in the UK annually, but is still very significant as it amounts to 27 million tonnes of "waste" each year.

More detailed and up-to-date data are available specifically for Household Waste, so we can break down this huge figure to understand what can be done to reduce it. Total UK Household Waste arisings weighed in at 27.3 million tonnes in 2016 and 26,411 million tonnes in 2018[3] (a 3% reduction) and there were 25.5 million households in the UK,[4] so this means the average household produced just over one tonne of "waste" each year.

But trying to pin down what actually makes up Household Waste is not easy. Just imagine your own "waste" bin. What is in it will not be the same as that of your neighbour and will change throughout the year. Every household is different, depending on the number of people in the household, their ages, their interests and their lifestyles.

The only way to establish representative data on the contents of our recycling and "waste" bins is to undertake "waste" surveys, which involve hand sorting samples of Household Waste and weighing each category. As you can imagine, this is both time consuming and expensive, so is not undertaken very often. But some published data are available and I have used the best that I could find.

My main source for the composition of Household Waste is a report by WRAP, a registered charity whose vision is "a world in which resources are used sustainably". The report was published in January 2020, but the data it contains is for the UK in 2017.[5] Table 2.2 shows what was in the UK's Household Waste in 2017.

Just a point of clarification here, Table 2.2 (from WRAP) shows total Household Waste arisings in 2017 as 27.3 million tonnes. The Government figure for Waste from Households for 2017 is slightly different at 26.9 million tonnes (1% lower). In everything that follows I have used the figure of 27.3 million tonnes of Household Waste produced in 2017.

But as I said, trying to pin down what is actually in Household Waste is not easy. Table 2.2 is the results from a single survey; do the survey somewhere else or at another time and you'll get different results. For example, the figure for food "waste" in Table 2.2 is five million tonnes per annum, but the equivalent figure from another WRAP study[6] gives a figure of 6.6 million tonnes per annum for food "waste" from households. So, trying to be precise about how much Household Waste is generated in the UK each year and, more specifically, what is in it is very difficult. So, I have done the best I can and have based my analysis later in the book on the above figures. But this must be taken as a single source of data and one that is really a snapshot at a point in time. What we can do, however, is generalise from the specifics of this data and draw still valid conclusions based on this data.

Table 2.2 shows that over one third of all Household Waste comprised packaging, slightly less than one third was products and significantly more

TABLE 2.2 Household Waste composition in 2017

Material	Tonnes	Percentage of total
Paper	3,004,938	11%
Card	1,796,665	7%
Glass	1,799,644	7%
Ferrous metal packaging	336,732	1%
Non-ferrous metal packaging	220,399	1%
Dense plastic	1,118,611	4%
Plastic film	896,687	3%
Sub-total – packaging	**9,173,676**	**34%**
Other ferrous non-packaging	337,112	1%
Other non-ferrous (non-packaging)	120,160	0%
Other dense plastic non-packaging	494,762	2%
Waste electrical and electronic equipment (WEEE)	474,536	2%
Household batteries	15,570	0%
Textiles	1,312,290	5%
Wood	1,091,450	4%
Paints and varnishes	42,294	0%
Non-packaging glass	101,415	0%
Other household hazardous waste	49,613	0%
Miscellaneous non-combustible	1,518,073	6%
Miscellaneous combustible	1,948,782	7%
Sub-total – products	**7,506,057**	**28%**
Food "waste"	5,019,185	18%
Garden "waste"	4,691,827	17%
Other organic "waste"	603,092	2%
Sub-total – organic	**10,314,104**	**38%**
Fines*	158,032	1%
Other "wastes"	85,806	0%
Sub-total – residue	**243,838**	**1%**
TOTAL	**27,237,675****	**100%**

Notes:

This is the summation of kerbside, Household Waste Recycling Centre and bring data.

* "Fines" are very small particles of indeterminate composition that cannot be treated in any way.

** The equivalent figure given in Defra "UK Statistics on Waste", Table 1, March 2020 is 26,897,000 tonnes, a difference of 340,675 tonnes or 1%.

than one third was organics. The kerbside collection of Dry Recyclables is really the collection of packaging and since all organic materials either are or could be collected by kerbside, then just under three quarters of all Household Waste could be collected directly from households using appropriate segregation containers. This is why I have focused so much attention on kerbside collection. Kerbside collection directly from households is efficient and is relatively easy for householders to deal with.

However, the remaining 28% of Household Waste is made up of products of one kind or another and the collection and subsequent treatment of these no-longer wanted products is problematic. The great majority of WCAs do not currently include products in their kerbside collection of recyclables. The exceptions are a very few who include household batteries and small WEEE. I say more about how these unwanted products should be collected and treated in Chapter 6.

Notes

1   Defra, "UK Statistics on Waste" Methodology, March 2020, p. 15. Waste from Households includes bulky waste and "Other Household Waste" taken to HWRCs.
2   You can see the impact here of Pre-Consumer Waste which comprises much of the category "Commercial and Industrial" and which is nearly 40% larger than Household Waste.
3   Defra, "UK Statistics on Waste", 19 March 2020, Table 1, p. 3.
4   Office for National Statistics, "Families and Households in the UK: 2019", section 1: Main Points, November 2019.
5   WRAP, "National Household Waste Composition 2017", Table 7, January 2020.
6   WRAP, "Food surplus and waste in the UK – key facts", October 2021, p. 5.

3

KEY DEFINITIONS AND CONCEPTS

Having talked about how we generate unwanted materials through our consumption, I'd now like to turn to how we currently manage those unwanted materials. Figure 3.1 illustrates the three positive options, what I call the Three Rs (all will become clear at the end of this chapter), together with the negative option of disposal.

In a moment I'll explain how we currently manage Household Waste and then talk about how we might improve this in the future but, before I do, I need to share with you some definitions so you will understand what I'm talking about. There are two categories of definition that I will be using and these refer to:

- Household Waste collection; and
- Household Waste treatment.

Household Waste collection definitions

Kerbside collection

The term "kerbside collection" means householders have to put their wheeled bin or recycling containers at the edge of their property on the designated collection day (it's called kerbside, because refuse used to be collected from wherever the dustbin was located, usually at the back of the house). With the introduction of wheeled bins, householders are required to wheel their bin to the kerbside on collection days, to reduce the amount of time taken

DOI: 10.4324/9781003504757-4

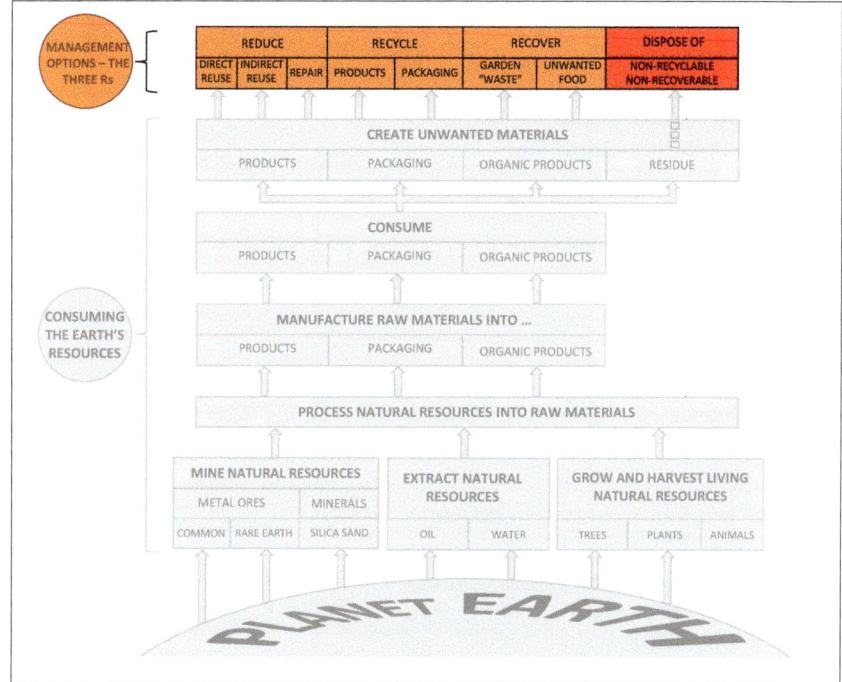

FIGURE 3.1   Options for the management of unwanted materials.

by the crews to collect and empty the bins and so save money for the Waste Collection Authority (your local council) and thus you as a Council Tax payer.

Drop-off

The alternative to kerbside collection is what is called Drop-Off or "bring schemes" where householders take their unwanted materials to specific sites where there are collection containers (called "banks") to collect separated materials. Examples are paper banks, bottle banks and can banks in public car parks, supermarket car parks and at Household Waste Recycling Centres.

Household Waste Recycling Centres

What used to be called Civic Amenity Sites are now called Household Waste Recycling Centres ("HWRCs"). These are sites provided by the local Waste Disposal Authority, not the Waste Collection Authority, where householders take separated materials for recycling or disposal. Please see the section "The players" on pp. 46–47 for an explanation of local authority roles. These sites allow householders to bring a very wide range of segregated materials, for example:

- paper and cardboard;
- glass and plastic bottles; and
- metal drink and food cans.

These are the so-called Dry Recyclables (see below) that are normally included in a kerbside collection scheme. In addition, HWRCs also usually accept segregated:

- household batteries;
- Waste Electronic and Electrical Equipment (WEEE), from mobile phones to fridges;
- computers, monitors and televisions and related computer equipment such as printers and printer cartridges;
- DIY waste, rubble and building materials;
- flat glass (as opposed to glass bottles);
- light bulbs and fluorescent light tubes;
- garden "waste" (but not food "waste");
- gas bottles;
- household and garden chemicals (for disposal, not recycling);
- oil, both cooking oil and engine oil;
- scrap metal and wood;
- clothing, textiles and shoes; and in some cases
- spectacles and books.

Each Waste Disposal Authority has its own list of accepted materials.

Dry Recyclables

When talking about Household Waste kerbside recycling schemes, there is a minimum range of materials that are collected and which are called Dry Recyclables. These are:

- paper;
- cardboard;
- glass containers (bottles and jars, but not window glass, glassware or broken glass);
- metal food and drinks cans (steel and aluminium);
- aluminium foil;
- plastic bottles; plus
- some hard plastic containers, i.e. food packaging, such as tubs and trays.

Some Waste Collection Authorities also collect textiles and shoes and WEEE in their kerbside collections, but these are in the minority.

The above list comprises the materials that are normally collected in a kerbside recycling collection scheme and apart from some paper, these items **are all packaging**. So, the collection of Dry Recyclables is primarily about recycling packaging.

Frustratingly, I often see terms printed on packaging describing one or more of the materials used in the product or packaging as "recyclable", but I know these materials are unlikely to be recycled because the Waste Collection Authorities don't collect them; they are not on the standard list of Dry Recyclables. As I've said, if they are not on the above list, they won't be collected. What the producer who prints this on the packaging means is that the materials are recyclable **in theory**, but in practice they won't be (please see my discussion on packaging labelling in Chapter 8).

A good example of this is plastic film used in packaging. Polyethylene (PE) film can be successfully reprocessed, but it is not cheap to do so and it needs to be separated from other polymers, something that is best done by the householder if they can recognise whether a particular film product, such as a plastic bag, is made from PE or not.[1] I have yet to find one Waste Collection Authority that collects plastic film in its kerbside scheme.

But I digress. I think we need a new and very specific definition of the term "recyclable".

A clear definition of what "recyclable" means

I would like to see a much clearer definition of how the term "recyclable" is used. I would like "recyclable" to only be used for materials that:

- are collected at scale:
  - for products this means take-back schemes must exist for the unwanted product via retailers (see Chapter 6); and
  - for packaging, the materials must be collected by all Waste Collection Authorities in their kerbside recycling collection schemes;
- can be separated into individual materials:
  - for products, a dismantling infrastructure must exist for the particular product; and
  - for packaging, this must be capable of being sorted into individual materials in a MRF; and
- are reprocessed to become the equivalent of virgin raw materials, economically and at scale.

In addition, to attain the levels of recycling that are potentially achievable, I advocate that all households should be provided with a single collection container (a wheeled bin to facilitate kerbside collection) into which they place all mixed Dry Recyclables. This is called "co-mingled collection" and I will say more about it in Chapter 10.

Household Waste treatment definitions and the Waste Hierarchy

There is a widely used concept called the Waste Hierarchy that was formally introduced by the European Commission in 2008,[2] although it had been talked about since 1990. The Waste Hierarchy sets out the order in which "waste" should ideally be treated, starting with the most environmentally acceptable method.

In simple terms, the Waste Hierarchy is described as:

- reduce;
- reuse;
- recycle;
- recover; and
- dispose of.

So, what impact has the Waste Hierarchy had on how we treat Household Waste? Table 3.1 shows how all collected Household Waste was treated in England from 2015 to 2019.

The Waste Hierarchy was introduced formally in 2008, but appears to have had very little impact on how much we discard or how we treat our Household Waste. The percentage of Household Waste recycled or recovered (as organic material) remained static over this five-year period at about 43%.

The only significant change that this table does show is a marked reduction in landfilling and a very big increase in EfW incineration (I say more about this in Chapter 5, but this change had nothing to do with the Waste Hierarchy and everything to do with costs). The message from Table 3.1 is that in 2019 we continued to generate just as much Household Waste as we did five years previously and we were doing little to change how we managed materials more sustainably when we no longer wanted them.

Whilst no-one would argue with the logic of the Waste Hierarchy as such, its fundamental flaw is that it treats everything as "waste". As I have and will argue, what householders throw out is not "waste" at all; it is by and large materials that the householder simply no longer wants. It mainly comprises valuable resources that should be recognised as such and treated accordingly.

TABLE 3.1 All local authority collected Household Waste (data for England only) ('000 tonnes)

Household Waste treatment method	2015		2016		2017		2018		2019		Change 2015–19	
Landfill	4,367	20%	3,578	16%	2,813	13%	2,373	11%	1,873	8%	(2,494)	-57%
Recycling and organic recovery	9,414	42%	9,735	43%	9,509	42%	9,409	43%	9,453	43%	39	0%
EfW incineration	7,877	35%	8,809	39%	9,496	42%	9,649	44%	10,043	46%	2,166	27%
Other	568	3%	647	3%	618	3%	602	3%	704	3%	136	24%
TOTAL	22,226	100%	22,769	100%	22,436	100%	22,033	100%	22,073	100%	(153)	-1%

Note: Department for Environment, Food and Rural Affairs, "Statistics on waste managed by local authorities in England in 2019/20", 3 March 2021. Author combination of Table 1, p. 5 and Table 2, p. 4.

Our performance to date

So how are we doing today?

As I've said, in 2017 the UK generated 27.3 million tonnes of Household Waste (if you're paying attention you'll have noticed this doesn't match the tonnage given in Table 3.1, but Table 3.1 is for England only and here I'm talking about the UK as a whole). How this "waste" was treated is summarised in Table 3.2. Of the total, 45% was apparently recycled.[3]

This shows the UK met the EU target for the recycling of Household Waste in 2018, but actually we're not doing very well at achieving what could be possible.

We generated 27.3 million tonnes of Household Waste in the UK in 2017, an average of 400kg per person, which is about six times our own average body weight.[4] That's how much we're each throwing away every year.

But, the apparent good news is that we are recycling 12.3 million tonnes of this "waste" each year (a Government reported recycling rate of 45%).

I say apparent, because of this 45% recycling rate, nearly half (19%) was organic material, mainly garden "waste" and as I will say below, this isn't recycling. But semantics aside, looking back to Table 2.2 in Chapter 2, 38% of Household Waste is potentially recoverable organic material, so we actually recovered only half of the organic material available.

Also, recovering garden "waste" is relatively easy; many people do it at home on their compost heap. Even doing it on an industrial scale is relatively easy, but only produces what is called a "soil conditioner". It's not the compost you buy in heavy plastic sacks at the garden centre to nurture and grow your tiny seedlings or potted plants. Soil conditioner has a much lower nutrient content than commercially produced "compost" and so is not suitable for growing seeds and cuttings. It is good for improving the organic content of soil, but it's relatively low in value, both environmentally and economically. Yes, recovering garden "waste" as soil conditioner is a good thing, but it's not

TABLE 3.2 How Household Waste was treated in 2017

Material	2017
Total reported as recycled	45%
- of which food "waste" recovery	2%
- of which garden "waste" recovery	17%
Total organic recovery	19%
- so material actually recycled	26%
Energy recovery and disposal	55%

Note: Department for Environment, Food and Rural Affairs, WFH_England_data_201819.xlsx spreadsheet, tab "WfH_Calendar", www.gov.uk, 2019, Table 2 Recycling.

recycling, because it can only happen once and doesn't return the unwanted material to its original state.

Thus, nearly half of what we apparently "recycled" in 2017 was easy to do, low grade recovery, not recycling. Only 26% of the 45% was Dry Recyclables.[5] Yet from my analysis presented in Chapter 11, up to 39% of Household Waste could potentially be recycled as Dry Recyclables. This means that only 67% of the available Dry Recyclables were recycled. So, the Government stated "Recycling Rate" of 45% is not so impressive when weighed against a potential combined recycling and organic recovery rate of 73%, i.e. it's only about 60% of what could be possible (I go into much more detail about what I think is possible in Chapter 11). So, there is a lot more that could be done. The truth is, we've done the easy part, we really have to change if we are to achieve the potential that is possible and really make a difference.

Can it be done? Can we change our ways? Yes, I believe it can and we can, but it will require changes both in how we each behave and in the infrastructure needed to deal with the materials that we no longer want. This will require a co-ordinated set of actions from: producers; retailers; you and me as consumers and householders; local councils; the waste management industry; and the UK Governments. But first, everyone needs to accept there is a problem and that the problem is urgent.

None of the changes required is so fundamental that it will be impossible to achieve. Most of what I will advocate in Chapter 10 is incremental change, not a radical overhaul of the way we live.

Revisiting the Waste Hierarchy

Throughout this book I will use the following treatment definitions:

- **reduce**: avoiding materials becoming unwanted through reduced consumption, reuse or repair:
  - reduced consumption: buy less and buy better (I'll say more later);
  - reuse:
    - direct reuse: using a product or packaging again for exactly the same purpose for which it was originally produced (this particularly applies to packaging, such as returnable/refillable bottles);
    - indirect reuse: using a product or packaging for a different, but still valuable purpose (this again particularly applies to packaging, for example glass jars or plastic boxes being used for home storage, cardboard boxes being used more than once by householders);

- repair: a form of reduction achieved by mending or replacing broken or damaged components to return a product to its original state, so it can continue to be used for the purpose for which it was originally manufactured, thereby not becoming "waste";

- **recycle**: turning unwanted materials (by reprocessing) into equivalent virgin raw material that can be used in the same way as virgin material, i.e. there is no degradation in the physical properties of the recycled material; ideally, the number of times a material can be recycled should be infinite, but as we shall see, there are limitations to the number of times some materials can be recycled;

- **recover**: turning unwanted materials into alternative materials or energy or both, giving them one further life, as opposed to recycling which is in theory, infinite:

  - organic recovery: converting unwanted organic material (principally food and garden "waste") into a useable solid component (soil conditioner) and combustible gas that can be burned to produce energy;[6]
  - energy recovery: incinerating materials under controlled conditions to generate heat and from this electricity (called Energy from Waste (EfW) recovery); and if all else fails

- **disposal**: final disposal to landfill of anything that is left after the above steps have been taken.

Direct and indirect reuse

As I said above, direct reuse is when a product or packaging is used again for exactly the same purpose for which it was originally produced, such as refillable glass bottles (and I'm struggling to think of another example apart from plastic "bags-for-life"), although some producers are introducing refill packs for household products such as cleaning products, in the form of pouches to refill the original heavier plastic bottle, thus reducing, but not eliminating, the amount of plastic being used. A better approach is for retailers to offer products in bulk containers from which customers can directly refill their empty containers. A few shops are offering this service, but many more could and I think should, do so.

But I'd also like to make a comment here on the concept of indirect reuse. Many people advocate reusing items for new purposes, to give them a second life. Miller and Aldridge have an interesting take on this in their book *Why Shrink Wrap a Cucumber?*[7] where they say "For re-use to work, however, it must make the purchase of another pack unnecessary. Simply making a bird-feeder out of plastic bottles is not a useful contribution to the environment; it does nothing practical to save on packaging and the bottles will eventually end in the waste stream".

This is harsh, but is it true? Yes, in that making a bird-feeder out of an old plastic bottle does not prevent a new plastic bottle being created for the packaging of the original drinks product. Direct reuse does do this, for example with a glass milk bottle. And once the bird-feeder has reached the end of its useful life (which it inevitably will), the unwanted plastic bottle will still need to be managed. Provided the bottle is recycled, you could argue that such indirect reuse is giving the plastic bottle a second, albeit temporary, but nevertheless useful life, before being recycled. However, the important point here is that using the unwanted plastic bottle as a bird-feeder saves the materials and energy needed to manufacture a non-bottle bird-feeder. So instead of displacing a virgin drinks bottle through direct reuse, this indirect reuse has displaced an alternative, virgin bird-feeder. So, I don't agree that such an example "is not a useful contribution to the environment", provided the indirectly reused item is properly recycled at the end of its life.

The general point I would like to make is that indirect reuse is a good thing, **provided that** such reuse does not prevent the reused item being managed in the way it would have been managed if it had not been indirectly reused, for example if it would have been recycled. Some types of indirect reuse actually prevent the recycling of the reused item at the end of its second life, because the materials are combined with others in a way that renders them non-recyclable. Such an approach is not a good thing. I'll give you two examples, both involving plastics.

The first example is turning unwanted mixed plastics into street furniture, such as benches and fence posts. The second is using unwanted polyethylene in road re-surfacing (I talk about both of these in Chapter 5 under down cycling). Both of these treatment approaches give the unwanted plastics a second life, but in doing so, prevent them from being recycled. As importantly, there is a high risk that when these products become unwanted, their further treatment could lead to the release of micro-plastics into the environment (this is particularly true of so-called, plastic roads).

I say more about direct and indirect reuse in Chapter 5.

Recycling vs. recovery

In the above text I have been quite specific above about the terms "recycle" and "recover". **They are very different**. True recycling should be a virtually infinite process, turning unwanted materials back into their original form that is indistinguishable from virgin material, again and again. Recovery is a one-off process that results in the recovered material being quite different from the original material and therefore no longer suitable for its original use.

With energy recovery (or EfW incineration to call a spade a spade), we burn the materials to heat water to produce steam, which is then used to produce electricity. By burning Household Waste we are not recovering materials we are recovering energy and not much at that (see Chapter 5).

The treatment of organic "waste"

But what about organic material treatment? We take food and garden "waste" and convert it into a low grade organic material (a soil conditioner) and from food "waste" we also produce gas that can be burned to produce electricity. So, is organic "waste" treatment recycling or recovery and does this matter? Yes it does and I'll explain why.

If the recovered material, the soil conditioner, is added back to the soil in which new plants are grown, this could be viewed as recycling, but I don't think it is. We are breaking down organic products (plants) into their constituent parts and then using these elements to **help** produce new products (new plants). But this is not the same as recycling where we turn unwanted materials directly into the same (but this time wanted) raw materials.

But if the soil conditioner is not used to grow new plants, for example when it is used on landfill sites as daily cover for newly deposited Household Waste (see Chapter 5), then the process is one of recovery, not recycling.

In addition, the energy produced from anaerobic digestion is clearly not recycling as we are recovering energy not materials, in a similar way to how EfW incineration does (as explained in Chapter 5). So anaerobic digestion and composting are both recovery, not recycling.

This issue matters, because apart from the need to be precise, it is important to understand how our Household Waste management performance is measured and reported.

Recycling and recovery are two very different things, but historically the European Commission and our own Governments have bundled recycling and organic recovery together and talk about single "recycling" targets and performance as if recycling and organic recovery are the same thing; they are not. So, I prefer to stick to the term "organic material recovery", because it emphasises this difference in treatment.

A new name to replace the "Waste Hierarchy"

I think we need a different name for the "Waste Hierarchy". Given that we are talking about materials and how we should manage them for maximum environmental benefit, I think we should talk about the Materials Management

Hierarchy or perhaps the Materials Hierarchy for short, which should comprise the following methods of treatment in decreasing order of attractiveness:

- **reduce**: avoiding materials becoming unwanted through reduced consumption, reuse[8] and repair;
- **recycle**: turning unwanted materials into the equivalent of virgin raw materials that can be used in the same way as virgin materials; and
- **recover**: turning unwanted materials into alternative materials or electricity or both, giving them **one** further life either via organic recovery or EfW incineration.

I'll call these the Three Rs (no not reading, writing and arithmetic, important as these are); but the Three Rs is simpler, clearer and logical.

If we cannot treat a material by one of the Three Rs, then the only alternative we have is **disposal**, i.e. the final, controlled disposal to landfill of anything that is left after the above Three Rs options have been exhausted.

A personal example of the Three Rs in practice

I was recently clearing out accumulated "stuff" and came across a large, scruffy bag that contained a three-person camping tent that my daughter used to take to music festivals. She no longer wanted it and neither did I. We didn't know what condition the tent was in; she thought it might have a rip in it and some of the poles might be missing or damaged. I didn't want to spend the time unpacking it to find out its condition, so the easiest thing to do would have been to take it to my local HWRC to put it in the disposal skip. A bad example of out of sight, out of mind, just chuck it because that's easy.

Instead, I put it on my local Freecycle site and made it clear that the tent's condition was unknown, but of course I was giving it away. Ten people replied, saying they wanted it! The chap who picked it up had repaired a similar tent before; his wife sewed up a tear in it and he bought on eBay some new poles that were needed, so he was happy to do this again to end up with a very good tent that had hardly been used.

This is an example of Reduction, through Repair, leading to Reuse. All it took was a little effort on my part (taking photographs and listing the tent on Freecycle) and for him to do what would be a relatively easy repair job.

This took the treatment method from the bottom of the Materials Hierarchy to the very top, from Disposal to Reduction. All it takes is a change in mindset and a little expertise; from just chuck it to repair it. That's why I say, "Don't ditch when down-sizing", make a little effort to preserve unwanted products. And that's the key point. We as consumers have to make a little more effort at the end of a product's life. Don't just ditch it, think about what others might do with what you no longer want.

The problem of focusing on recycling

The Three Rs rank the preferred methods of treatment in decreasing order of acceptability, which is determined by the degree of environmental benefit that the treatment method delivers. So:

- reduction is the highest preferred method of treatment, because it avoids materials becoming unwanted, either through reduced consumption (which is the highest possible aim) or reuse and repair, both of which preserve the materials and the energy that have gone into the product's or packaging's production;
- recycling is the next best option because it at least preserves the materials used within a product or packaging, but not the energy, costs and other inputs associated with the product's or packaging's manufacture which are lost; and
- recovery is one step further down, because both the materials, energy, costs and other inputs consumed in the product's or packaging's manufacture are lost and only a very small quantity of down-graded material and limited energy are recovered.

The Materials Hierarchy puts the highest priority on reduction, but there seems to be very little attention being given to this treatment method, I think because it is seen as being much harder to achieve than recycling. I'll talk in detail in Chapter 5 about how improvements in reduction could be achieved, but for now I want to shine a light on the issue that **it is all too easy to skip over reduction as the highest priority approach and instead focus on recycling**. Ignoring reduction as the top priority needs to stop.

Too many people, from Government to householders, think the solution to our problems with Household Waste Management is to recycle more. Their logic is that we'll just recycle everything we no longer want, then what's the problem?

But as I have already said, our currently reported recycling rate of 45% is in fact only 27%, a pitifully low number. Yes, we could increase the rate of recycling, but let's just take a step back.

Recycling is only a partial solution, because whilst materials are preserved, all of the energy, costs and other inputs used in the original product or packaging manufacture are lost. Yes, materials are preserved, but so much is lost.

The reason recycling is seen as being easier to achieve than reduction is that, in theory, unwanted materials can be collected from households, sorted if necessary and then reprocessed to provide raw materials that can be used to manufacture new products and packaging. This sounds simple and is the basis for the much-hyped Circular Economy (see Chapter 6). But this thinking is flawed for two reasons.

First, when most people talk about kerbside recycling, they are talking about schemes that only collect Dry Recyclables, which comprise predominantly simple packaging. All **products**, for example WEEE and clothing, are excluded from virtually all kerbside recycling schemes, as is multi-material packaging such as drinks cartons. So, the scope for expanding the recycling of Household Waste to include products, for example, is extremely limited.

Second, and this is a very common misconception, there are very few materials capable of being reprocessed indefinitely without degrading. The ones that are capable are: metals, glass and potentially some plastics. But paper, cardboard, textiles and wood are not, as I will show in Chapter 5. These materials can only be reprocessed a handful of times, before they become too degraded to be of further use. So, again the scope for expanding the recycling of many materials in a truly circular manner is very limited.

But, another very important point I want to make is that when we talk about recycling, we are talking about reprocessing **individual** materials. Any unwanted product or packaging has to be reduced to separate, individual materials before these can be reprocessed. The more complex a product or packaging, the harder this is to achieve. So, unless products and complex packaging are **designed** for material separation and the infrastructure necessary to achieve this separation is put in place, then recycling will have a limited role to play in improving the treatment of Household Waste. Recycling is not the panacea many people seem to think it is.

The players

The Supply Chain

People often talk about the Supply Chain when discussing the production, distribution and retailing of goods. By this, they are referring to all the stages that take raw materials, through manufacturing and distribution to become products and packaging that are purchased by consumers either in physical shops or online. This is a linear, one-way process that ends with the consumer. I have illustrated this Supply Chain in Figure 3.2.

FIGURE 3.2   The traditional Supply Chain.

But as I have been saying, we have to think about what happens to the products and packaging when consumers no longer want them. The Supply Chain definition needs to be extended beyond the consumer and ideally become circular, rather than linear. I say more about this in Chapter 6, but for now I will simply say that in an ideal world, recyclable materials would be collected, sorted and reprocessed so the producers can use the materials again to make new products and packaging. But as we will see later, this is the ideal and things don't yet really work this way.

Who does what

There are a number of players involved in the Supply Chain and beyond, so I'd like to start by looking at who does what, then we can go on to see how each one needs to change to create a sustainable Circular Supply Chain.

The producers

Producers make things. These are either:

- products (clothes, electronic goods, food and drink, etc.); or
- packaging (in which products are wrapped to protect and promote them and which usually gets discarded pretty quickly).

The retailers

Retailers buy products from producers and sell them to us, the consumer, for a profit. It's as simple as that.

Retailers can have physical stores, such as supermarkets or small shops or can promote and sell online.

But a very important point here is that large retailers can have a very significant influence over how producers create and package their products.

The consumers

This is you and me. We buy stuff, we use/consume stuff and we chuck it out when we're done with it.

Local authorities

There are two systems of local government in the UK:

- the so-called two-tier system of local government comprising a number of District or Borough Councils within a geographical county, together with a single County Council; and

products have first to be collected and then dismantled, in order to produce single streams of materials, suitable for reprocessing.

So how is this currently done? In terms of collection, some larger products are already collected at HWRCs, for example fridges, freezers and television sets; smaller items are mixed together in a "Small Electricals" skip. But what if you can't get your item to an HWRC?

Most WCAs operate what they call a Bulky "Waste" collection service, where the householder pays to have bigger items collected by the WCA from their home, but what about smaller products?

Some retailers will take back your old product when you go into their shop or when they deliver, say, a replacement fridge, freezer or television set. This is usually on a one-for-one basis, so what are you going to do with your drawer-full of old mobile phones and accessories? I'll say more about this later.

Dry Recyclables collection

The collection of Dry Recyclables is much simpler than product collection and can be done in two ways:

- kerbside collection from households; and
- drop-0 collection by householders.

Kerbside recycling collection

A very few WCAs operate a co-mingled collection service, where all the Dry Recyclable materials are mixed together. Co mingled collection works well from two points of view:

- it is as easy as possible for the householder to use as there is no separation required; and
- existing refuse collection vehicles can be used to collect the recyclables, thus maximising WCA vehicle efficiency, particularly as the recyclables are compacted in the collection vehicle, thus maximising its payload.

However, because the recyclable materials are collected mixed, they must be sent first to a MRF for sorting, before they can be sent on for reprocessing.

The co-mingled collection of recyclables and subsequent sorting in a MRF has been shown to work well. My own experience in Leeds back in the 1980s showed that mixing recyclable materials in a modern compactor collection vehicle had no detrimental effect on the collected materials and they could be successfully separated at the MRF.

FIGURE 4.1 A standard refuse compactor collection vehicle. Photograph courtesy of Houston PR on behalf of Biffa.

There was an initial concern that glass containers would break or that food waste would contaminate materials such as paper. However, experience showed that the level of glass breakage is surprisingly low, due to the cushioning effect of the other co-mingled materials.

Provided there is a separate food "waste" kerbside collection scheme and that householders are encouraged to rinse their discarded food containers, such as tins, the issue of food "waste" contamination is also minimal.

This is why I advocate co-mingled recyclables collection, in wheeled bins, using existing compactor collection vehicles (see Figure 4.1 for an example of a modern compactor collection vehicle). Please see Chapter 10 under the sub-heading Recycling for a more detailed discussion of this.

Kerbside collection

Different WCAs currently provide different collection containers for Dry Recyclables and require differing degrees of separation by the householder. No WCA requires the householder to separate recyclable materials into individual materials in separate containers, but the majority require some degree of separation, for example into mixed paper and glass in one container and all the other recyclables in another, perhaps with cardboard in a third container. Containers can be a wheeled bin, 55-litre boxes with lids, or heavy-duty colour-coded bags.

Where there is some consistency between WCAs is that in almost all kerbside recycling collection schemes, there is a minimum range of Dry Recyclable materials that are collected:

- paper;
- cardboard;
- glass containers (bottles and jars, but not window glass or glassware);
- metal food and drinks cans (steel and aluminium);
- aluminium foil;
- plastic bottles; and
- some hard plastic containers, i.e. food packaging, such as tubs and trays.

The number of different materials that can be collected is often limited by the number of compartments on the collection vehicle (if collection is source separated). This means some materials that could be collected aren't, for example textiles, household batteries, small WEEE and plastic film. Glass isn't separated into colours but is collected as a mixture of colours.

Source separated recyclables are collected in collection vehicles that are different to a normal refuse compactor collection vehicle. Source separated collection vehicles are divided into compartments into which the recyclable materials are hand sorted by the collection crew (see Figure 4.2). The collection of recyclables is either carried out on a fortnightly rota with non-recyclable

FIGURE 4.2 A source separated collection vehicle. Photograph courtesy of Cheltenham Borough Council/Ubico Ltd.

Household Waste collected one week and Dry Recyclables (and separated garden "waste") the next, with food "waste" collected every week.

But the problem with this approach (apart from asking the householder to separate out their unwanted materials) is a very practical one, in that the compartments in the collection vehicle fill up at different rates. I recently spoke to the driver of one of these vehicles who told me:

- his vehicle's cardboard compartment was full, even though he was only part-way through his round, so he had had to call in a second collection vehicle, just to collect cardboard; and
- then his glass compartment had filled up, so his crew were leaving all the glass to be collected by a **third** vehicle that he had called in!

The collection vehicle and crew efficiency was thus dire, due to the fixed nature of the collection compartments in the vehicle. This problem is inevitable as the mix of Dry Recyclable materials changes over time and even week-by-week. For example, there has been an explosion of cardboard usage as a result of online shopping deliveries, hence the problem with the cardboard in my example above.

The solution to this is co-mingled collection, using standard wheeled bins and a standard collection compactor vehicle, as explained in Chapter 10.

Organics – kerbside collection

Garden "waste"

Many WCAs operate a garden "waste" collection service. Some operate this all the year round, others only during the peak gardening months (March to October). Some WCAs charge for this service, others do not.

Garden "waste" may also be deposited free of charge by householders at HWRCs. Collected garden "waste" is taken to large-scale composting facilities (see Chapter 5).

Food "waste"

Many, but not by any means all, WCAs collect segregated food "waste" from the kerbside, but this is set to change (see Chapter 5). A small 23-litre caddy is provided for food "waste" which is collected weekly and sent for anaerobic digestion (see Chapter 5).

Household Waste for disposal

This is the traditional "dustbin" collection. This is now almost entirely carried out using wheeled bins, although some WCAs require householders to put out black refuse sacks (usually where streets are too narrow for a collection vehicle to access the houses).

A large, wheeled bin (typically 240 litres) is normally provided for non-recyclable "waste" and separate, smaller containers for recyclables and, sometimes, segregated food "waste". This sends the message to the householder that more "waste" for disposal is expected than recyclables, which is the wrong message to send.

Multiple collection containers

One of my concerns about the current Household Waste collection arrangements for Dry Recyclables is that WCAs are providing householders with too many collection containers and that different WCAs are giving their residents different types of collection containers, so it can all be a bit confusing for the householder, particularly when they go on holiday within the UK.

Examples of different types of kerbside containers for Dry Recyclables, food "waste" and garden "waste" are shown in Figure 4.3 (no household is expected to use all of these containers).

As an example, my WCA gives its residents: one 240-litre wheeled bin for "waste" for disposal; two 55-litre boxes for recyclables: one for mixed glass and paper and one for co-mingled metal cans, aluminium foil and

FIGURE 4.3   Examples of kerbside collection containers.

rigid plastics; a reuseable bag for cardboard; and one 23-litre caddy for food "waste". The food "waste" caddy is quite adequate for all times of the year except Christmas when the extra vegetable peelings and the turkey carcass won't fit in, so I keep some of the vegetable peelings for collection the next week.

Given that most WCAs practise what is called "source separation" collection, householders are given several different recyclables containers and are asked to segregate their recyclables into different (usually colour-coded) containers. As another example, one WCA I researched gives each household:

- a blue plastic box for paper;
- a red plastic box for co-mingled metal cans, aerosols, aluminium foil and plastics (but not plastic film);
- a green plastic box for glass, cardboard and Tetra Pak type cartons;
- a 23-litre "caddy" for food waste;
- reuseable bags for garden waste; plus
- householders are asked to put out: small electrical items, textiles and shoes, and household batteries, each in separate carrier bags.

It is interesting that many WCAs ask householders to set out some items in carrier bags, but with the advent of the carrier bag 10p levy, householders have far fewer carrier bags nowadays and what do they do when they have no carrier bags?

For larger properties, storing all these containers, plus a 240-litre wheeled bin for non-recyclable Household Waste is much less of a problem than for smaller properties, for example terraced houses. But for smaller properties, storing all these containers is a real headache. There is also a risk that householders will put recyclable items into the wrong container, either because they make a mistake or just can't be bothered (let's be honest, some people will not take the trouble required).

The alternative to source separated collection is what is called "co-mingled" collection where all the recyclables are mixed together in one container, usually a wheeled bin. I say more about this in Chapter 5.

What happens if we don't kerbside recycle?

One of the problems with giving householders so many collection containers and asking them to separate their Household Waste into so many different categories is that some people get confused and make mistakes (for example, by putting Household Waste that should be set out for disposal in with the Dry Recyclables), or worse, just can't be bothered when faced with what they see as a complicated system.

When householders do not source separate their Dry Recyclables correctly, the actions taken by the WCA vary, but two approaches are common:

- The householder receives a warning letter, followed by further warning letters if the non-participation or incorrect setting out continues, culminating in a visit from an "education officer" who tries to explain and convince the householder how to recycle correctly. In the event of further non-compliance, the recycling service is withdrawn from that household to avoid further contamination (but this doesn't help increase recycling).
- The householder is fined up to £100 for each breach of the recycling guidelines.

The good news is that such enforcement action is rare as most householders are keen to participate in recycling. However, one area where householders are less co-operative is in food "waste" collection. One WCA that provides a food collection service told me they only get 50% participation from householders; for others the participation rate is much higher.

But enforcement action is as a result of the failure of the WCA to educate the offending householder how to recycle and a failure to convince the householder why recycling is important. This is why we need to make the recycling collection arrangements simple and easy to use (KISS) and we need to educate everyone in what is required and why (for more on this, please see Chapter 12).

Drop-off

Drop-off is how Household Waste recycling started, with the very earliest examples being glass bottle banks. Whilst drop-off remains a form of recycling collection, it is reducing as kerbside schemes have been introduced, but my key concern is that it is both environmentally and economically inefficient, because:

- householders have to transport their recyclable materials to the collection points which are relatively few and far between (some supermarkets provide limited drop-off banks so householders can drop-off their recyclables when they do their shopping), so this is an inefficient use of transport;
- not all householders are able to take their recyclables to drop-off points, particularly if they do not have a car, so that not all the recyclables that could be collected are collected, which led to the development of kerbside collection schemes;
- collection banks fill up at unpredictable rates so either they fill up before being emptied leaving no space for further materials (but some people

still leave their recyclables on the ground next to a full drop-off bank), or if they are emptied on a regular basis, on some occasions they will not be full when emptied, making the collection inefficient; and

- the collection banks have to be collected by specialist vehicles that have to make long trips to collect a bank and deliver it to the point where it is emptied, adding further environmental and economic collection costs.

Free-standing recycling banks can still be found in some public car parks and at larger supermarkets. Such banks are usually limited to newspapers, cardboard, glass bottles, plastic bottles and maybe textiles. But cash-strapped WCAs are increasingly removing drop-off banks from all locations apart from Recycling Centres.

Whilst the use of drop-off banks is generally declining, they still have a role to collect those materials that are **not** included in a local kerbside collection scheme and this is most visible at HWRCs.

The contribution from recycling drop-off

Whilst it is always good when materials are collected for recycling, drop-off collection will only ever collect a very small fraction of the materials that could be recycled, for the reasons given above, particularly because not everyone has a car to enable them to get to a car park collection bank or HWRC.

My analysis suggests that the contribution from bring collection represents only 4% of current Dry Recyclables collection for recycling.[1]

Household Waste Recycling Centres

Whilst drop-off banks are inefficient, they are the method of collection used at HWRCs. A wide range of materials is collected for recycling, wider than the range included in kerbside collection because the materials do not have to be sorted after collection as they are deposited separately on site.

HWRCs facilitate the collection of non-Dry Recyclable items such as: textiles, shoes, books, cooking oil, batteries, paint, engine oil, scrap metal, building rubble, WEEE items, light bulbs, as well as Dry Recyclables collected through the local kerbside scheme, such as paper, glass, cans, foil and plastic bottles.

One of the biggest quantities of material collected is garden "waste", deposited by people who do not want to pay for a kerbside garden "waste" collection service. HWRCs do not accept food "waste", but do take Household Waste for disposal, both small items as would be placed in a disposal kerbside bin and larger items like tyres, furniture, DIY surplus materials and other bulky "waste" that is too big to fit into a wheeled bin.

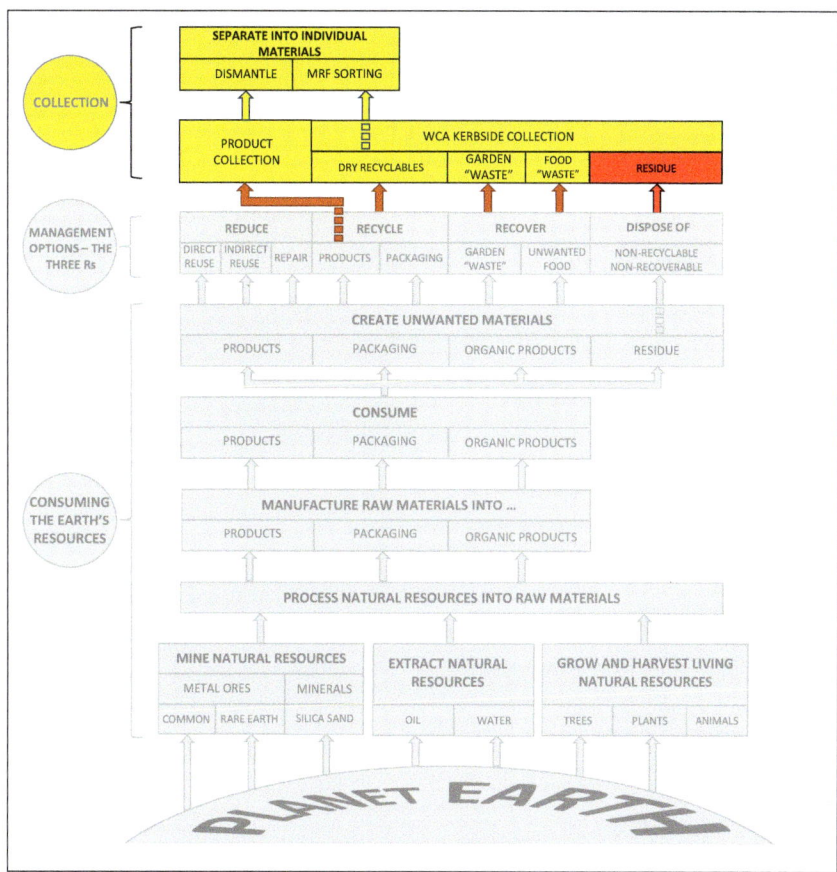

FIGURE 1.4 The ideal approach to Household Waste collection.

The ideal approach to Household Waste collection

In Figure 4.4 I have summarised how I think different materials should in future be collected. I discuss this more in Chapter 10.

Note

1 WRAP, prepared by Eunomia Research Consulting Ltd, "National Household Waste Composition 2017", January 2020. Author analysis of Table 7.

5

HOW OUR HOUSEHOLD WASTE IS TREATED

The Three Rs

If we go back to the Materials Hierarchy, there are three ways of treating unwanted materials (the Three Rs):

- reduce (including reuse and repair);
- recycle; and
- recover.

Any material that cannot be treated by one of these methods has to be disposed of.

Let's start with the first of the Three Rs by looking at how we can reduce Household Waste.

Reducing Household Waste

Consume less

It goes without saying that if we consumed less, we wouldn't generate so much Household Waste, so this has to be the starting point for reduction. But, simply saying to people that they should consume less isn't going to make it happen; it's a complicated subject, one which I explore in Chapters 6 and 7.

For now, I'd like to focus on the two practical ways that we can reduce Household Waste, which are reuse and repair.

DOI: 10.4324/9781003504757-6

Reuse

As I said in Chapter 3, there are two kinds of reuse: direct packaging reuse and indirect product and packaging reuse.

Direct packaging reuse

Direct packaging reuse is when packaging is used again for exactly the same purpose for which it was originally produced.

The large-scale direct reuse of packaging has declined over time (glass beer and soft drinks bottles used to be reused, driven by a bottle deposit system, but no longer are). If we're honest, direct packaging reuse is currently limited to just two items: glass milk bottles and "bags-for-life". Yes, there are small-scale, local schemes involving glass wine bottle and beer growler refilling and reuse, but I'm struggling to think of any other examples of any scale.

There are also some limited opportunities for indirect packaging reuse, such as when consumers use empty bags or containers to hold food products that are sold loose, but at the moment this is pretty small beer.

Indirect product and packaging reuse

I define indirect reuse to mean three things (and this applies more to products, but can apply to some packaging):

- packaging: filling empty packaging containers (bags or tubs) in-store with products (usually food products) that are sold loose;
- products:
  - someone else reusing what I no longer want or need; and
  - in keeping with the above, me buying used or second-hand; or
- reusing packaging or a product for a new purpose, other than its original purpose.

I say more about this in Chapter 10.

Repair

Like direct packaging reuse, I have struggled to think of large-scale repair systems for household products. Two that come to mind are shoe and clothing repairing and clock and watch repairing, but can't think of any others, apart from car repairing.

The thing that clocks, watches and cars have in common and which encourages us to repair them when they fail, is that they are relatively expensive to replace, so we're willing to pay to have repair work done. I suggest shoes and clothes are different. Yes, they are relatively expensive to replace, but they are also simple to repair and we are often quite attached to them. Thus, a repair infrastructure has grown up to satisfy this consumer demand for the repair of all of these products.

But this is a pretty short list. So many other consumer products could be repaired, but generally aren't, for three reasons:

- too many products are not designed to be repaired (either to make it cheaper to manufacture them or to encourage consumers to buy a new replacement rather than to repair a broken one);
- the infrastructure necessary to repair products just doesn't exist (where do you take electrical products to be repaired?); or
- as I said earlier, we as consumers have been programmed to always want new products, rather than preserving what we already have.

So, my message is simple, we need to repair more products rather than discarding them.

Treatment other than reduce

So that's my take on "reduce". Any method that comes below "reduce" needs the collected materials to be treated in some way.

As a starting point, Table 5.1 shows how Household Waste was treated in England in 2019.

Data on how much Household Waste is incinerated or landfilled are not available, so I have taken advice from Defra's Statistics Team (and my thanks to them for their help) and have derived these figures for England based on 2019 data.[1]

The surprising conclusion that comes from this analysis is how much of our Household Waste is incinerated (45%) and how little is landfilled (only 8% of the total). What is worrying is that the number of EfW incinerators being

TABLE 5.1 How UK Household Waste was treated in England in 2019

Treatment method	'000 tonnes	Percentage of total
Recycling and organic recovery	9,453	43%
EfW recovery	10,043	45%
Landfill	1,873	8%
Other	**704**	**3%**
TOTAL	**22,074**	**100%**

built in the UK is set to increase dramatically and could limit any increase in recycling or organic recovery (I say more about this in Chapter 10).

Having talked about Household Waste reduction as the first of the Three Rs, the next two Rs (Recycling and Recovery) involve some form of post-consumer treatment of the collected materials, so I'd like to say a few words about treatment cycles, before moving onto recycling and recovery.

Treatment cycles

I rather like David Attenborough's idea that we are dealing with two different cycles which he calls the biological cycle and the technical cycle. He said "Anything that is naturally biodegradable – food, wood, clothes made from natural fibres – is part of a biological cycle. Anything that is not – plastics, synthetics, metals – is involved in a technical cycle. The raw material in both cycles – the carbon or titanium, for example – are elements that need to be reused".[2]

As a former engineer, I'd like to label them slightly differently, whilst still meaning the same thing. I'm going to talk about the mechanical treatment cycle and the organic treatment cycle. These cycles are illustrated in Figure 5.1.

The blue arrows in Figure 5.1 show how reprocessing creates valuable outputs. The red and orange arrows demonstrate that that all forms of reprocessing create residues that have to be treated either by EfW incineration or landfilling.

Mechanical materials and the mechanical treatment cycle

What follows is a general discussion of how Dry Recyclables are treated. For a more detailed description of how each of the main materials, for example paper or glass, is reprocessed, please see Annex I.

David Attenborough talked about "Anything that is not [biological] – plastics, synthetics, metals – is involved in a technical cycle", or what I have called a mechanical cycle.

Mechanical treatment applies to those materials for which recycling must be the primary means of end-of-life treatment. But, the materials this applies to are surprisingly limited:

- paper and cardboard;
- glass;
- metals; and
- plastics.

As explained below and in Annex 1, some materials, specifically paper and cardboard, textiles and wood are can only be recycled a limited number of

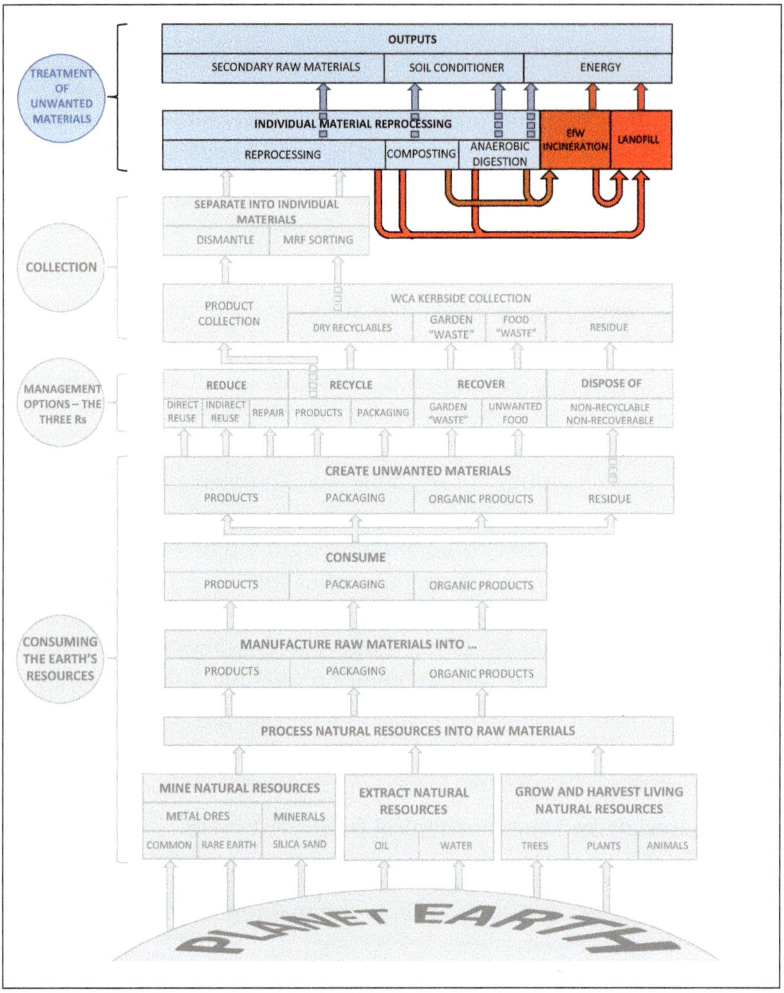

FIGURE 5.1 The treatment of unwanted materials.

times, then they have to be recovered, usually as energy by EfW incineration. This is summarised in Table 5.2.

Glass and metals are, in theory, infinitely recyclable, as are single polymer plastics and can be turned back into virgin-equivalent raw materials at a much reduced cost, both in environmental and financial terms, compared with the production of virgin materials. They can also be turned back into the products from which they were originally recycled, so glass can be turned back into glass bottles and jars and metals back into aluminium or steel cans, a true closed loop, circular economy.

TABLE 5.2 The number of times materials are, in theory, recyclable

Material	Theoretical number of times material can be recycled
Glass bottles and jars	Infinite
Aluminium and steel cans and aluminium foil	Infinite
Plastic bottles and rigid containers	Infinite
Plastic film	Infinite
Paper and cardboard	Five to seven times
Textiles	One to three times
Untreated wood	One to three times

But, I am concerned that too many people think that just increasing recycling is **the** answer. As I hope I have shown and will show, recycling is a very large part of the answer, but it is only one of the things that we need to do differently, in order to address our Household Waste problems.

Recycling

As I have said, many, many people think recycling is the answer to our Household Waste problems and they diligently separate out their unwanted materials week in week out, with the hope and expectation that these materials will be sorted and reprocessed and turned back into useable raw materials. And whilst this is usually the case, there is a real danger that consumers see recycling as a panacea that absolves them of having to moderate their consumption. This view is one of "well, I wanted to buy it, but now I don't want it, but it's okay, because it'll get recycled, so no harm done". Such an attitude is wrong on so many levels.

We must minimise the creation of unwanted materials through Household Waste reduction (the easiest form of which is: if you're not sure, don't buy it); then we should reuse and repair as much as we can and only then does recycling come into play.

And it's not just a question of householders collecting more materials for recycling. There are changes needed to the way we are asked to separate materials for recycling and to the way these materials are collected and sorted.

So, let's dig into recycling which comprises three stages:

- collection;
- sorting; and
- reprocessing.

I've already written about how recyclable materials are currently collected and say how this should change in future in Chapter 6, so I'll move straight onto sorting.

Sorting of recyclables

No householder is asked to separate their recyclables into individual materials for kerbside collection. All materials are collected mixed to some extent and so need subsequent sorting. Some WCAs carry out simple sorting manually at the kerbside, into a vehicle that has a number of different compartments (see Figure 4.2 in Chapter 4), whilst a relatively low number of others take co-mingled materials to a Materials Recovery Facility (MRF) for a mixture of manual and automated separation.

Obviously, the contents of drop-off banks do not need sorting as they are collected as single materials, but they do need checking for contaminants.

Co-mingled material sorting in MRFs

Since the early 1990s, MRF technology has developed to the extent that now glass bottles and jars can be automatically colour sorted and plastic bottles made from certain polymers can be automatically segregated into their specific plastic polymer types. But not all polymers can be automatically detected so, logically, these polymers should not be used to make plastic bottles, for example, PVC.

The whole thrust of MRF technology development has been to make the sorting process as automated as possible. This is to minimise costs, maximise the quality of the separation and in recognition of the fact that few people want to stand sorting other people's recyclables for seven hours a day. Believe me, I've done it and there are certainly more attractive jobs.

Thus, every WCA should have access to a local MRF to which it can deliver co-mingled recyclables for manual and automated separation.

Recyclable material reprocessing

Once mixed recyclables have been sorted into individual material streams, they are bulked up (either loose, as in the case of glass or compacted into bales, as is the case for paper, cardboard and plastic bottles). The materials are then shipped in bulk to different reprocessors who convert the recyclables into useable raw materials.

Design for recycling

In order to maximise the recycling of Household Waste, we have to recognise two very important facts:

- first, not all materials can be easily recycled; and
- second, if a material cannot be easily recycled then in my view it shouldn't be used by producers in their products or packaging.

This is another radical statement and goes against what most producers currently do. Producers, not surprisingly, use whatever material suits their purposes and product costings. So, we have a multiplicity of materials being used in durable products and packaging, many of which are not currently recycled and to be honest never will be because it is too complicated and expensive to separate and reprocess them.

I get quite annoyed whenever I see the label "Not currently recycled" on packaging, for example, as shown in Figure 5.2. What really gets to me is the arrogance and lack of responsibility shown by the producer.

They have decided to use a material that cannot be recycled and appear to expect someone else to solve the recycling problems. This is the antithesis of the concept of Producer Responsibility (see below).[3] This statement should never appear on packaging; if it isn't currently recycled, this is because it isn't currently recyclable and in all probability never will be. So why is the producer using this material? Only because it suits their purposes and hang the consequences. As I say, this makes me quite annoyed.

FIGURE 5.2   Examples of bad packaging labelling.

If we accept my proposed definition of "recyclable" materials, then producers should only use those materials that are truly recyclable to produce their products and packaging. If the infrastructure does not exist to collect products or packaging and separate these into individual materials at scale, it should not be termed "recyclable". Okay, in theory it might be recyclable and Tom Szaky argues that any material is recyclable in theory,[4] even used nappies and cigarette butts, but if the practical constraints of collection schemes and material separation mean a material cannot be successfully collected and then economically reprocessed, then it should not be termed "recyclable".

We need producers to design products and packaging that are not only easy for retailers and WCAs to collect for recycling, but which will actually be recycled. This means that the products and packaging have to be made from the limited range of materials that it is possible to easily recycle (see Chapter 10).

In addition, it has to be possible for householders to easily separate different, individual materials as part of recycling collection. A good example of how this can be done is large yoghurt pots comprised of a thin plastic pot (too thin on its own to be strong enough for handling and transport), with a cardboard sleeve wrapped around the plastic pot to give it strength and stability. The cardboard sleeve even has a zip printed on it, with perforations, making it obvious and easy to remove the cardboard sleeve before putting the plastic pot and cardboard sleeve separately into the recycling collection container. If one producer can do this, why can't they all?

Another good example is Jiffy bags, which have been specifically designed for recycling. The inner polythene plastic bubble wrap, which forms the protective layer of the Jiffy bag, can be easily separated from the outer paper envelope. Whilst this does require the householder to physically separate the two layers, this is easily done and clear instructions are printed on the paper envelope. The two layers are then ready to enter the two different reprocessing routes, one for plastic film and one for paper.

A third example of design for recycling that I came across recently was a meat pie in an aluminium foil tray, packaged within a simple printed cardboard outer box, with no plastic window. The box carried an On-Pack Recycling Label or OPRL (see Chapter 8) which simply said "Recycle", no explanation needed, because both cardboard and aluminium are recyclable.

Producer Responsibility

Just to explain the concept of Producer Responsibility, this obliges producers of packaging to take responsibility for the environmental impact of their packaging. The Producer Responsibility Obligations (Packaging Waste) Regulations 1997 (replaced by the Producer Responsibility Obligations (Packaging Waste) Regulations 2007) enacted the 1994 EU Waste Packaging

Directive and obligated producers to pay a proportion of the cost of the recovery and recycling of their packaging through a system of Packaging Recovery Notes (PRNs). Registered reprocessors of recyclable materials issue PRNs and packaging producers and retailers can buy these to fulfil their obligations under these regulations. It doesn't mean producers have to actually recycle or recover their packaging, they just have to pay someone else to do it for them.

This concept has been expanded into the idea of Extended Producer Responsibility which requires producers to add all of the environmental costs associated with a product throughout the product life cycle, including its end-of-life treatment, to the market price of that product. Thus, the consumer pays the costs of treatment at the end of the product's/packaging's life, which to my mind is quite correct.

Extended Producer Responsibility for packaging was due to be introduced by the UK Governments in April 2024. However, in July 2023 this was pushed back by a year to avoid the inevitable consumer price increases adding to the cost of living crisis.[5]

The three ways of reprocessing recyclables: Up, down and around

Sorted Dry Recyclables and unwanted products can be reprocessed in three very different ways:

- circular/closed loop recycling;
- so-called "upcycling"; or
- down-cycling.

Circular/closed loop recycling

This is what most people think of when they think of recycling.

In closed loop recycling, an unwanted material is turned back into a clean, useable form of the original material. For instance: aluminium drinks cans are recycled into clean aluminium ingots that can be used to make new drinks cans; collected and sorted "waste" paper can be reprocessed to produce new paper, as can cardboard; glass bottles and jars are converted into a raw material called "cullet" (essentially broken up glass, which is used to make new glass bottles and jars); and of course steel food cans can be recycled into new raw steel that can be used for almost any steel product, including cans (it's the original scrap metal industry, after all).

But, recycling of course doesn't come free. It requires considerable energy, generates CO_2 emissions and will inevitably result in some unwanted materials being rejected from the reprocessing plant. For example, plastic windows from envelopes and cardboard packaging for food are "waste" products from paper and cardboard reprocessing, as are paper labels from glass bottle and metal can recycling. So, when we see recycling rates quoted for different materials or packaging types, we need to understand that these usually represent the percentage of material collected for recycling, not the actual percentage recycled (which will be less).

But also, not all collected materials can be fully recycled. For example, only 80% of paper that is sent for reprocessing comes out as new paper. Paper is made from paper fibres (which come originally from the trees from which paper is made) and each time paper is recycled, some of these fibres break and become too short to be of use and are "washed out" during reprocessing. This "waste" paper sludge has to be disposed of. And as I've already said, there are also "waste" products to be dealt with, such as the ink that is removed from the previously used paper and plastic windows from window envelopes and cardboard packaging.

So, recycling doesn't come free, but for the key Dry Recyclables, it is infinitely preferable to recycle these materials than to produce them from scratch. Making glass, aluminium, steel, paper and plastic are massively energy intensive, whereas the energy needed to reprocess these materials is a fraction of the initial energy investments in these materials.

You won't be surprised to read that reprocessing aluminium drinks cans saves 95% of the energy required to mine, process and smelt the aluminium ore to produce aluminium can stock. Recycling steel saves 70% of the energy needed to produce virgin steel, and recycling paper and cardboard saves 40% of the energy needed to pulp trees and convert them into virgin paper.[6] And don't forget the two benefits associated with closed loop recycling: fewer trees need to be grown and felled for paper and cardboard production, freeing up land for other purposes; and huge areas of Australia will not have to be desecrated to mine bauxite (the ore from which aluminium is produced) or sand for glass manufacture.

As well as not consuming precious, finite resources and the environmental desecration that goes with this, the unwanted materials will not have to be incinerated or landfilled to dispose of them. So, I think that's a win-win-win for recycling.

Upcycling

As well as closed loop recycling, an increasingly popular term is "upcycling". I get what people mean by this, which is to turn one product into another, in order to increase the value of the original product. For example, in his book *Revolution in a Bottle* which sets out the history and philosophy of the recycling company TerraCycle,[7] Tom Szaky tells the story of how his company TerraCycle realised that large oak wine barrels were seen by the Californian wine industry as "waste" after only one use (further use apparently tainted the wine). At the same time, he was talking to a retailer who wanted alternatives to plastic water butts and plastic composters. Putting two and two together, TerraCycle bought and "upcycled" unwanted oak wine barrels into attractive wooden water butts and composters. This turning of a "waste" product into an alternative product with a higher value is what Tom calls "upcycling". I've seen other examples such as: turning old heritage hand operated machinery into "retro" or "vintage" light fittings; or repainting old furniture with chalk paint to make it look more attractive.

Tom's example of converting unwanted wine barrels into water butts and composters is a clear case of adding significant value to an otherwise unwanted "waste" product. But I question some of the lesser examples, such as simply painting old furniture with chalk paint to make the old product look new or different or trendy, so someone will be tempted to buy it. The danger is that this actually devalues the original product, by turning it into a trendy, sometimes not very good, potentially short-lived impulse purchase, which will soon be discarded. If the original product had been properly repaired and restored, it could have had much greater value and would potentially last and be valued for much longer.

I think there are three issues here.

First, we're talking about complete products being repurposed in some way. The product is not broken down into its constituent materials and then these materials individually reprocessed to produce new raw materials; the product stays intact as a product. So, there is no recycling taking place. In fact, I would argue that this is a form of reuse, albeit indirect reuse.

Second, to be successful, this indirect reuse has to add real value to the new product for it to be successful, otherwise why would anyone pay for it?

Third, if this form of treatment is to have any significant impact on the management of our Household Waste, it has to take place at scale. "Upcycling" a few pieces of furniture is not going to make a major difference. So, I would say that upcycling is a niche activity, not a fundamental solution to the treatment of Household Waste.

Down-cycling

Down-cycling produces a product that whilst it has an intrinsic value is worth less than the original product; it diminishes a material's value.

For example, mixed plastic waste can be turned into useful products like street furniture (bollards, benches and fence posts), provided it has enough polyethylene (polythene to you and me) in the mixture of unwanted plastics. The polyethylene acts like a glue or matrix to hold all the other chips of waste plastic together (polyethylene melts at 180°C–270°C)[8] so by heating the waste plastic and squeezing it through a nozzle (this is called extruding) a continuous strip of recycled plastic material can be produced, with different profiles or cross-sections for different purposes. Such recycled plastic products are usually made in black, to disguise the different waste plastics within them. This limits the uses to which such materials can be put, but it is better than burning or landfilling them. But what about when they reach the end of their life? Can they be recycled? In theory yes, but in practice?

Another example of using polyethylene as a glue is what are being described as "plastic roads", for example that from MacRebur, where "waste plastics are added to reduce the volume of bitumen required in an asphalt mix".[9]

But again, what happens (a) to the plastic as the road wears and (b) when the road is resurfaced and the plastic infused tarmac is ground up?

Mechanical treatment: Conclusions

For mechanical materials, such as glass, metals and potentially plastics, the primary form of treatment should be recycling, turning the discarded materials back into the equivalent of virgin raw materials.

For those materials such as metals and plastics for which there is a finite supply of raw materials (the Earth can only provide so much aluminium and iron ore and oil for plastics production), we either need to replace them with sustainable materials or manage them as precious materials that must be recycled. But given their abundance, their extraction and recycling would appear an acceptable approach, provided the recycling rates are very high.

Organic materials and the organic treatment cycle

Sir David Attenborough argues that the organic cycle involves "Anything that is naturally biodegradable – food, wood, clothes made from natural fibres ..." This raises an interesting question: does our use of organic materials involve them in a circular economy or a simple linear (one time only) process? The answer, not surprisingly, is that it depends on the material.

For some materials, such as food and garden plants, the process is a linear one: grow; process (food); consume (food); treat unwanted materials by incineration or landfill, unless the organic "waste" is treated by anaerobic digestion or composting. This linear cycle is shown as (A) in Figure 5.3.

For other organic materials, for example wood and natural fibre textiles, their life is a mixture of one or more cycles then a linear path to disposal. This is shown at (B) in Figure 5.3. The life of such products is:

- the materials are grown and processed to make them suitable for manufacture;
- the materials are manufactured into consumer products;

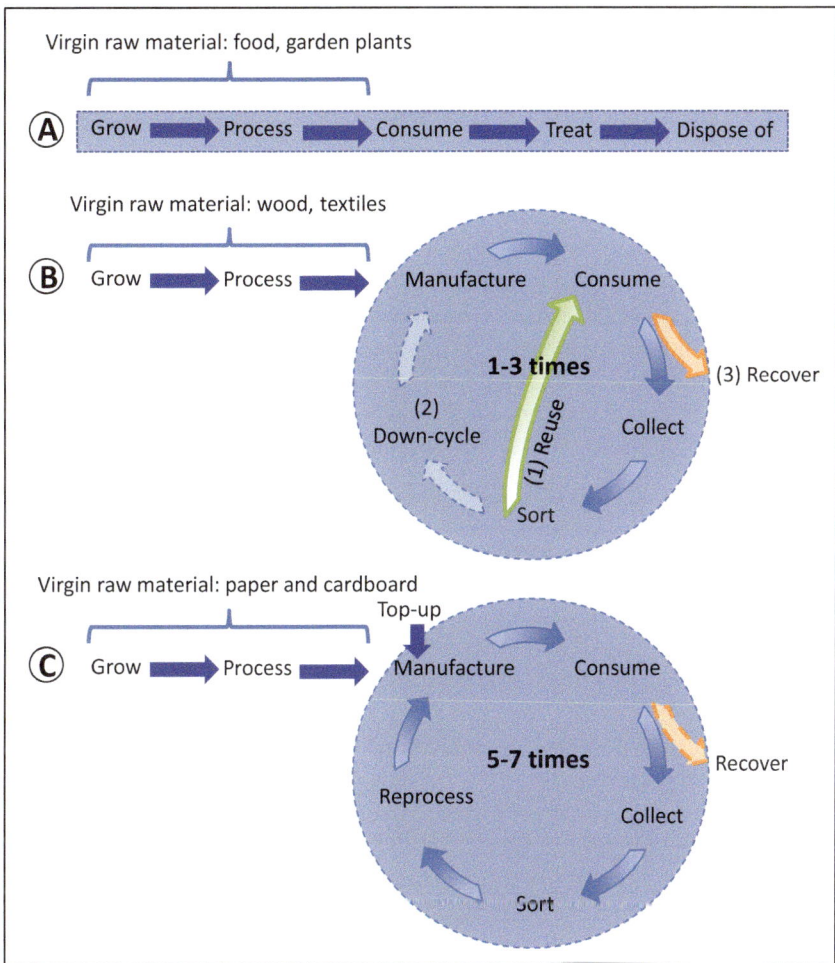

FIGURE 5.3 The treatment of organic materials.

- we consume these products until we no longer want them and they are (or should be) collected and sorted into either:
  - products suitable for reuse, e.g. clothes (shown as (1) in Figure 5.3 (B)); or
  - materials to be down-cycled, e.g. already reused wood or clothes and other textiles (shown as (2) in Figure 5.3 (B)) into less valuable, but still useful raw materials;
- for products that are to be reused, these are simply prepared for a second life;

- in the case of down-cycling, the materials are remanufactured into consumer products, e.g. textiles that are no longer suitable for reuse as clothes can be down-cycled into rags for use as cleaning products;
- once we no longer want the reused or down-cycled products, these are (or should be) collected and sorted, either for down-cycling (in the case of reused products) or recovery; and
- finally, once the materials have been through reuse and possibly a down-cycle, they are recovered (shown as (3) in Figure 5.3(B)).

These materials, such as clean wood and natural fibre textiles, can thus have up to three lives if collected, sorted and processed correctly.

Taking a closer look at no-longer-wanted wood, provided it has not been treated with paints or varnishes, it can be used more than once. For example, there are charitable schemes that collect and resell discarded, good quality wood to people who want to use it to make things.[10] This is shown as (B1) in Figure 5.3. But this can only really happen once or at the most twice, before the wood becomes unattractive for further use or such use becomes impractical. The wood could potentially then be down-cycled (provided it is still clean and untreated), for example by being chipped for use as animal bedding or being chipped and made into chipboard-type products which give the original wood a new life, although this is quite unusual. Following such a third life the wood is finally rejected and should be sent for either organic or energy recovery, i.e. composting or EfW incineration. So, option B could give the raw materials between one and three lives.

Finally, let's look at paper and cardboard. As I said above, whenever paper and card are reprocessed, a proportion of the fibres in the paper and card break and become too short to form part of the recycled paper and card; they are then discarded from the process. Paper and cardboard can only be recycled five to seven times before the fibres become too stiff and too short to make new paper. So, whilst paper and cardboard manufacture and use is a circular process, it can only happen a limited number of times, before a proportion (typically 20%) of the paper/cardboard input stream is rejected.

As Figure 5.3 shows, the organic treatment cycle can only ever be partially circular and for some materials, in particular organic materials such as food and garden plants, the treatment process has to be linear.

What we must do is ensure we make the best use of these organic materials, once they have been through any processing, rather than simply burning them or burying them in a landfill site. And to follow the theme of a biological cycle, they should be returned to the earth in a form that will enhance the earth, rather than damaging it: as a soil conditioner to improve the quality of the soil.

There are two treatment processes for doing this: anaerobic digestion (for food "waste") and composting (for garden "waste").

The treatment of food "waste" by anaerobic digestion

The Anaerobic Digestion Strategy and Action Plan published by Defra and The Department of Energy and Climate Change in 2011 described anaerobic digestion as:

> Anaerobic digestion (AD) is a natural process in which micro-organisms break down organic matter, in the absence of oxygen, into biogas (a mixture of carbon dioxide (CO_2) and methane) and digestate (a nitrogen-rich fertiliser). The biogas can be used directly in engines for Combined Heat and Power (CHP), burned to produce heat, or can be cleaned and used in the same way as natural gas or as a vehicle fuel. The digestate can be used as a renewable fertiliser or soil conditioner. Anaerobic Digestion (AD) is not a new technology, and has been widely applied in the UK for the treatment of sewage sludge for over 100 years. However, until quite recently it has not been used here for treating other waste or with purpose-grown crops.

What this description doesn't say is that anaerobic digestion must take place inside a sealed reactor vessel[11] (a relatively simple pressure vessel) to exclude oxygen and to capture the biogas that is generated, or that the CO_2 captured can be separated from the methane and used as a raw material.

If organic material such as food "waste" is buried in a landfill, it rots down and produces methane gas. Modern landfill sites capture some, but not all, of this methane and use it to generate electricity. However, some methane inevitably escapes into the atmosphere and contributes to global warming.

By processing food "waste" in a sealed anaerobic digester, all the methane produced is captured and is used to generate electricity or as an alternative to natural gas (it is known as Renewable Natural Gas). In addition, the solid residue from anaerobic digestion (known as "digestate" and representing 80% of the incoming food "waste" by weight) is used as a valuable fertiliser, thus reducing the amount of artificial fertilisers used on our fields.

Anaerobic digestion is thus a very good processing technique and increasingly WCAs are collecting food "waste" from households and sending it to small-scale anaerobic digestion plants for local processing and sale. Indeed, as farmers adopt anaerobic digestion as a way of treating their own animal "wastes", some are diversifying into offering anaerobic digestion capacity for the treatment of domestic and commercial food "waste".

Anaerobic digestion treatment facilities are thus increasingly available and can provide a local, relatively small-scale solution for the treatment of food "waste". This begs the question, if anaerobic digestion is such a good treatment method, why aren't all WCAs collecting food "waste"? If your WCA doesn't provide you with a food "waste" collection service, maybe you should ask them why not. In fact, the Environment Act 2021 requires all English WCAs to separately collect household food "waste" from 2023, bringing them into line with WCAs in Scotland, Wales and Northern Ireland,[12] so it's happening, but too slowly.

The treatment of garden "waste" by composting

Composting is a form of recovery, not recycling, because it is a once-only reprocessing opportunity that doesn't turn the inputs back into their original form. It does generate a product that has an economic value, but it cannot be seen as closed loop recycling, which can go on almost indefinitely.

In contrast to anaerobic digestion, composting is carried out in the presence of oxygen, typically in the open air. Please refer to Annex I for a description of the composting process.

Commercial composting produces good quality, sterile compost that doesn't contain active weed seeds, but does contain some of the nutrients plants need to grow.

Such recovered compost should not be confused with potting compost sold by garden centres. Potting compost has a very even and fine texture which is just what delicate seedlings need. It is also high in nutrients; however, these nutrients are soon depleted, which is why seedlings need to be potted on.

Recovered compost on the other hand contains a proportion of organic material which has yet to break down. This is a distinct advantage when adding it to garden soil as a soil conditioner, as unlike potting compost it will release nutrients over a long period of time as the material continues to break down. The fibrous nature of recovered compost is also good for conditioning soil, i.e. improving the soil texture, water retention capability and structure. It is apparently also excellent for attracting worms, which help to break it down further.

Like anaerobic digestion, the composting of domestic garden "waste" is a good treatment method, in that it removes organic material from EfW incineration or landfill and produces a useful product. Given that the breakdown of the organic material happens in the open air, little or no methane is produced, thus reducing the impact of the treatment of organic waste on global warming.

Such commercial composting is really just a large-scale version of what happens in a domestic compost heap but with the addition of regularly turning of the composting material and is operated at higher temperatures. Some so-called "compostable" packaging, such as corn-based "plastics", state on them that they are compostable, but neglect to say that for this to happen, the materials need to be commercially composted at these higher temperatures; often they won't compost down in a home compost heap.

Home composting and wormeries

Many keen gardeners create their own compost heaps onto which they put their garden "waste". Indeed, this has always been the traditional way of dealing with garden "waste". Domestic compost heaps take longer than commercial compost farms to break down organic materials, but are nonetheless very successful, if created and maintained correctly.

Householders are advised to not put food "waste" onto their compost heap as this can attract vermin. So, a useful adjunct to a compost heap is a home wormery, which is used to treat food "waste" in a closed (but not sealed) container. The wormery contains a specific species of worm (called brandlings: red, manure or tiger worms, not earth worms), which are introduced in large numbers. The worms break down the food "waste" into a friable soil conditioner and a liquid which can be captured and used as a liquid fertiliser.

Wormeries are not for everyone, indeed neither is home composting, and both should be regarded as niche, but nonetheless useful, treatment techniques. The advent of large-scale WCA food and garden "waste" collection schemes has really scaled up the treatment of organic Household Waste and made them a key part of the solution to treating elements of our Household Waste.

What happens to treated organic Household Waste?

The final stage of the treatment of organic materials, whether by anaerobic digestion or composting, is the loss of the physical material back to the earth, mimicking what happens in nature.

Is organic material treatment linear or circular? Whilst anaerobic digestion is a form of energy recovery, the materials are used only once before being returned to the earth. However, one could argue that if the digestate or compost is used to grow new, useful plants then the process is circular, so maybe organic recovery is circular in the right circumstances.

And if the process is linear, rather than a circular process, maybe it is not necessarily a bad thing. Provided we produce new organic materials (wood, paper, food, natural fibre textiles, etc.) in a sustainable way, then perhaps the above linear process is acceptable, but only if we extract the maximum environmental value at every stage.

Energy recovery

Energy from Waste incineration (EfW)

If mechanical products or packaging are unsuitable for recycling and if organic materials cannot be recovered, then the final option for their treatment that offers any environmental benefit is energy recovery in the form of EfW incineration; to be blunt, burning them, an increasingly popular option, at least with WDAs.

Table 3.1 in Chapter 3 shows how the landfilling of Household Waste in England declined over the five years from 2015 to 2019 (from 20% to 8% of total Household Waste) as EfW incineration grew from 35% to 46% (an increase of 27%).

The UK Government does not publish data on how Household Waste is treated, but data is available for England, which represents 85% of UK waste arisings, so this data is a reasonable proxy for the UK situation. And the published data is not for Household Waste, it is for all local authority collected "waste" (which includes some commercial "waste"), so I have had to interpret this data. Table 3.1 is thus an approximation, but it serves to make the point about the very significant rise in the use of incineration and the decline in landfill over time.

How an EfW incinerator works

Every EfW incinerator has the following basic elements:

- a waste reception hall, where the incoming "waste" is received;
- a combustion grate, where the "waste" is burned;
- a boiler surrounding the combustion chamber, comprising steel tubes through which water passes and turns to steam;
- a steam turbine that is driven by the steam from the boiler and which, through a linked generator, produces electricity;
- equipment to quench and scrub the combustion flue gases; and
- equipment to process the Incinerator Bottom Ash (IBA) which is what is left after combustion.

The IBA comprises a mixture of non-combustible materials such as ceramics, glass and metals and the solid residue from combustion, known as clinker. The IBA is processed to extract the metals which can then be recycled and the remaining IBA is graded to become low-grade construction aggregate. On average IBA represents about 20%[13] of the "waste" inputs to an EfW incinerator, in other words 80% of what goes in is converted to energy and gaseous emissions and the remaining 20% is recovered either as low-grade aggregate or as metals.

The gases that are discharged to the atmosphere after what is called flue gas "scrubbing" have to fall within strict limits imposed by the UK Environment Agency. Anyone living near to an existing or proposed EfW incinerator is naturally concerned about what is or might be emitted from the very tall chimney of the incinerator. We only have the operator's and regulator's word that these emissions are safe.

The chemicals that are used to scrub the flue gases and the particulate matter that is filtered out of the flue gases are called Air Pollution Control Residues (APCRs), which are classified as toxic waste. APCRs equate to a further 3.3% of the "waste" inputs and have to be landfilled as hazardous waste under strictly controlled conditions.

In addition, the energy efficiency of burning "waste" is questionable, being around 15%–27%[14] and which, of course, depends upon the calorific value of the "waste" inputs. The more paper, cardboard and, in particular, plastics we recycle, the lower the calorific value of the residual "waste". By way of comparison, figures from the World Coal Association, suggest that the average efficiency of coal-fired power plants around the world today is 33%, with modern state-of-the-art plants being able to achieve rates of 45%.[15]

There is also the fundamental issue that burning "waste" generates CO_2 at 250–600kg per input tonne.[16] A report in 2021 stated that "… the data that incinerator operators publish in their annual performance reports suggest their plants emit almost as much CO_2 per kilowatt hour … as a coal-fired power station".[17]

EfW incinerators represent a form of recovery as they generate electricity from Household Waste. But as I have said before, this is the lowest form of "waste" recovery as, apart from the recovery of metals and low-grade aggregates, all the materials that go into the plant are lost to future use.

UK EfW incineration capacity

Now I've always thought there is limited EfW incineration capacity in the UK, but it turns out this is far from the case.

There were 48 Energy from Waste (EfW) incinerators operating in the UK in 2019, with six more being commissioned and a further 12 under construction.[18] The combined capacity of the existing and being-commissioned incinerators was over 15 million tonnes per annum of municipal "waste" (that is Household Waste plus other "waste" of a similar nature that WCAs and WDAs collect). Many EfW incinerator operators also accepted significant quantities of Commercial and Industrial Waste (about 20% of total capacity).[19]

The combined electricity output from all UK EfW incinerators in 2019 was 6.7 GWh (representing approximately 2% of total UK electricity generation).[20]

But as Table 5.1 shows, a staggering 45% of Household Waste was incinerated in England in 2019 and the capacity to burn is increasing. This is worrying and I say more about it in Chapter 10 under the heading "The problem of excess EfW incineration capacity".

Landfill

If no other treatment option is available, we come to the very bottom of the Materials Hierarchy, that of landfill. Don't be mistaken, this is not a method of treatment, it is literally burying the problem of unwanted materials and managing the side effects of this action. It is also an abdication of our responsibility to manage the Earth's resources responsibly and sustainably.

Most people think that once you put materials into a landfill site, that's it; put it in the ground and forget about it and all will be well. And for past generations this was the only way to manage unwanted materials, which were simply seen as "waste". But putting unwanted materials into a landfill site is far from being a case of bury and forget.

Landfill sites are active reactors

A landfill site is not an inert thing. Landfill sites contain some materials that do not change once they are buried, for example glass and plastic. Other materials degrade very slowly, such as metals, textiles and wood, but other materials break down relatively quickly, particularly organic materials, such as food and garden "waste". Any breakdown of buried "waste" gives rise to two unwanted by-products that nowadays are actively managed by the landfill site operators (but in the past they were not, hence the problems experienced when building on old landfill sites, notably subsidence and landfill gas emissions).

The two by-products of landfilling are landfill gas (a mixture of methane and carbon dioxide) and "leachate" which is a potentially highly toxic liquid that comprises liquids that are tipped into the landfill site when it is active (for example paint and liquid household chemical products) or by-products of the degradation processes that go on within a landfill site.

Landfill gas, if not managed, contributes to the problems of global warming. Because there is no oxygen within a landfill site, the breakdown of organic material such as food and garden "waste" is anaerobic, which produces methane and carbon dioxide. Methane is about 23 times more damaging to the atmosphere than carbon dioxide[21] and is a major contributor to global warming.

Any new landfill site is more than just a hole in the ground. Yes, it starts as a hole, but nowadays the hole is lined either with clay or plastic or both, to seal it to prevent leachate escaping and polluting the local groundwater. Once the site has been lined, "waste" is deposited each day and spread out and compacted by specialist compactor vehicles. At the end of each day the newly deposited "waste" is covered by a layer of "landfill cover" which comprises low grade, inert solid "waste" such as building rubble, to provide an intermediate cover to prevent the "waste" being blown about by the wind and to limit access to it by the ubiquitous seagulls and other vermin.

In modern, more recently built landfill sites, as the "waste" is tipped into the site, a network of perforated pipes is laid within the "waste" to capture and collect the landfill gas as it is generated. This collected gas is then burned on site to generate electricity which is consumed on site and if there is enough, exported to the National Grid.

The Landfill Tax

Because landfill sites are active sources of pollution, even when no longer in use, tougher environmental management practices have caused landfill sites to become more expensive to operate. But landfilling was still cheaper than recycling until 1996 when the Government introduced the Landfill Tax[22] to artificially increase the cost of landfilling, in order to encourage greater recycling. The Landfill Tax was £7 per tonne of "waste"[23] when it was introduced and it has risen year-on-year. Today it stands at over £94 per tonne[24,25] and so has moved landfill from being the cheapest form of "waste" treatment to one of the most expensive (which was precisely why the tax was introduced). This has also had the effect of making recycling much more attractive to WCAs and WDAs from a purely financial point of view.

Future landfill capacity

Data on how much future landfill capacity exists in the UK is hard to come by, as some of this data is commercially sensitive. What follows is simply an estimate, based on official sources.

The remaining landfill capacity in England in 2019 was 370,992,455 cubic metres [26] Taking a density figure of 560kg/cu m for EWC code 20 03 99 "municipal wastes not otherwise specified",[27] this equates to 208 million tonnes of currently unused capacity. Table 2.1 (Chapter 2) shows the total "waste" generated each year in the UK is about 220 million tonnes; this suggests that there is a severe shortage of landfill capacity, which is one of the factors that has led to the explosion in EfW incineration capacity.

The loss of valuable materials

The reason landfilling is at the bottom of the Materials Hierarchy is because it results in the loss of valuable materials, as well as having significant environmental impacts. It is the opposite of a circular economy; it consigns valuable materials to what is essentially a grave and creates pollution problems that have to be managed. So, the less material that ends up in landfill the better.

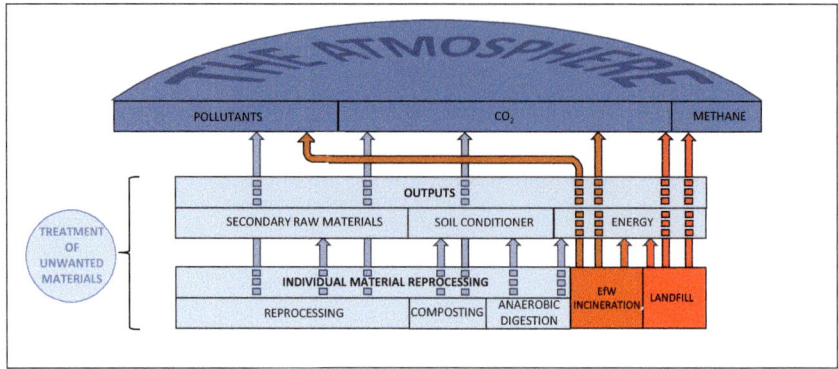

FIGURE 5.4 Atmospheric pollution from the treatment of unwanted materials.

Atmospheric pollution

But we don't only risk polluting the earth when landfilling; all "waste" treatment methods actually pollute the atmosphere as Figure 5.4 shows.

The relative costs of different treatment methods

Remember the Three Rs?

- reduce (including reuse and repair);
- recycle; and
- recover.

And if we can't treat a material by one of the Three Rs, then the only alternative is disposal to landfill.

How a WCA or WDA chooses which method to adopt often comes down to cost. To be fair, some local authorities give greater weight to environmental considerations than others, but in the harsh world of local authority funding, cost is often the deciding factor.

When I was first working in this field, landfill disposal was so cheap that it was very difficult for local authorities to implement recycling or recovery schemes as they were so much more expensive than landfill. This situation only changed when the Government introduced the Landfill Tax as a way of artificially increasing the cost of landfill, to make recycling and recovery more financially viable.

This is good in that it has made the costs of recycling and recovery much more competitive with landfill. However, it has had the unintended consequence of also making EfW incineration more competitive with landfill and in some cases, in terms of some materials such as plastics, it has made

FIGURE 6.2 The linear model of consumption.

them (which could be within minutes if we're talking about packaging, or years if we're talking about consumer goods such as furniture or a washing machine), the materials that make up the product or packaging are waiting for their next incarnation.

If we're lucky they may be:

- reused in their original form (such as a glass milk bottle);
- repaired if broken and then reused in their original form (for example, some electrical goods or shoes); or
- recycled as individual materials into valuable, reclaimed materials (think paper, glass, cardboard, aluminium and other metals) that can be used in new product or packaging manufacture, replacing virgin raw materials.

But, as Table 5.1 (in Chapter 5) shows, over half of what we no longer want either goes to an EfW incinerator or directly to landfill.

We are slowly but surely stripping our planet of irreplaceable raw materials like crude oil and aluminium (not to mention the rare earth metals used in electronic goods like mobile phones, electric car batteries and wind turbines) and polluting that same planet by discarding what we no longer want. You don't need me to tell you about the plastics crisis in our oceans and increasingly in our own food chains.

The challenge we face is how to turn our current linear consumption model into a circular one, where what is no longer wanted becomes the input into a new product cycle, rather than being discarded.

Now, there are people who think about how we manage our Household Waste once we've created it and how we can manage it more efficiently and effectively. But I think this is the wrong way to look at the problem. It is not a question of "how can we manage our 'waste' better?" but one of "how can we consume in a sustainable way, so we don't create 'waste' in the first place?" Ideally, we should all consume less, but I'm not naïve enough to expect this to happen in a big way. After all, our whole global economic model is based on consumption: producers make products; we buy and consume them; and they make more and we buy more. It's how the world works. But what should happen to these products when we no longer want them?

Circular consumption

We need a circular model of consumption that recognises that everything we consume comprises valuable resources that have to be managed in a sustainable way, specifically at the end of a product's useful life. So, when a product or its packaging is no longer wanted or needed, instead of simply disposing of it:

- if it has been designed to be reused, then we reuse it; or if not
- if it is a product and has been designed to be repaired, then we repair it and return it to its original consumer or sell it to a new consumer; or if not
- for both products and packaging, we harvest the valuable materials within it and feed them back into the production process as raw materials, which is essentially what recycling does.

This would essentially be the Three Rs in action and is illustrated in Figure 6.3, where Circular Consumption is highlighted by the light green arrows. Instead of an unsustainable, linear consumption model, we need to develop a sustainable, circular model that values resources. And we have to stop talking about "waste". David Attenborough summed this up rather well in his book *A Life on Our Planet* when he said:

> The key to the circular mindset is to imagine replacing the current take-make-use-discard model of production with one in which raw materials are thought of as nutrients that must be recycled, just as nutrients are in nature. It then becomes clear that we humans are engaged in two different cycles. Anything that is naturally biodegradable – food, wood, clothes made from natural fibres – is part of a biological cycle. Anything that is not – plastics, synthetics, metals – is involved in a technical cycle. The raw material in both cycles – the carbon or titanium, for example – are elements that need to be re-used. The cleverness comes in designing ways to do so.[1]

People have started talking about a "circular economy" and "circular economics".[2] This sounds great, but it's not that simple, certainly not yet. Because far too many of the products we buy and their associated packaging have not been designed for a circular economy, rather they are designed for disposal.

And very importantly, we can't expect to change to a more sustainable approach, without first redesigning many of the products we consume. This is a fundamental point. Products and packaging have to be "designed-with-the-end-in-mind". This will take time and the will of producers to do this.

That said, there are enough products available in the right form for us to start now. Most paper, cardboard, glass, metal and, to a lesser extent, plastic

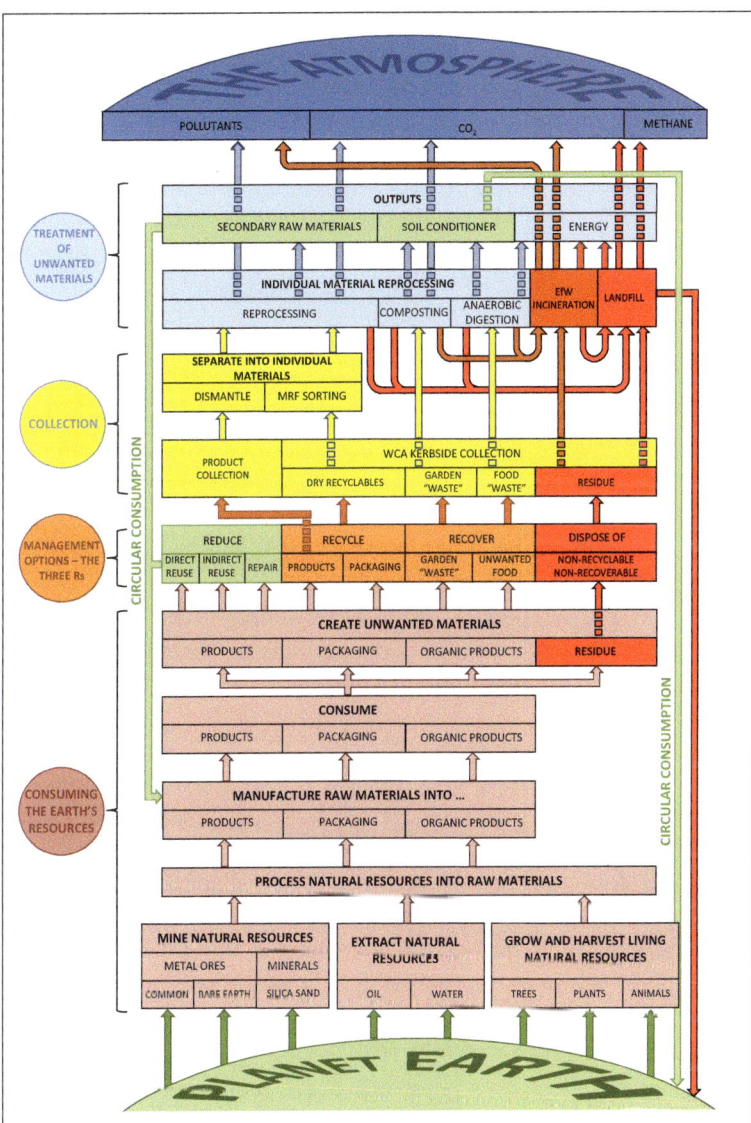

FIGURE 6.3   Circular Consumption.

packaging is suitable for recycling, which is a circular process. But we don't do enough of this for the reasons I have explained. But the real problem comes with **products**, where very few products are designed-with-the-end-in-mind, making it very hard to separate different materials, in order to recycle them. This is where the focus for change for producers needs to be.

The key message here is that producers need to redesign their products to suit a circular model of consumption. But then, and this is a key aspect of this new approach, the infrastructure has to be put in place to collect the unwanted products and packaging and to deliver them to specialist reprocessors. Not only does the infrastructure have to be in place, but consumers have to be willing to make the effort to use it. This will require a change in mindset, from "just chuck it" to "how do we reuse/repair/recycle this?" And the key to making this transition is the KISS principle. Unless we make it easy for householders to do the right thing, the majority simply won't.

Challenging the Supply Chain

In Chapter 3, I explained that what is today called the Supply Chain is all the stages that take raw materials through manufacturing and distribution to become products and packaging that are purchased from retailers by consumers. This is a linear, one-way process that ends with the consumer. This traditional Supply Chain has been optimised to deliver packaged products as efficiently as possible to the consumer.

I have also said that we need to extend this Supply Chain to become a Circular Supply Chain, to include the collection and treatment of unwanted materials once a consumer has finished with them. There are two ways this could be achieved:

- force the unwanted products and packaging back down the traditional Supply Chain, by retailers taking back unwanted products and packaging and sending them back to the original producer for treatment, effectively reversing the distribution channels (but this is not really viable for imported products); or
- extend the Supply Chain, beyond the consumer, to include the collection, sorting, separation and reprocessing of unwanted products and packaging, by involving all those players responsible for these activities, namely: WCAs, WDAs, the Waste Management industry and material reprocessors.

The first option is challenging, because the Supply Chain has been optimised for very efficient one-way distribution, but only as far as the consumer. As a product moves through the Supply Chain, it is handled by more and more players. A single producer will use multiple distributors, who serve multiple retailers, who sell to many, many customers. So, the Supply Chain disperses products as widely as possible, but very efficiently. Under this option, the Supply Chain would have to be made to work in reverse, whilst still delivering new products to consumers. This is why I say it would be challenging.

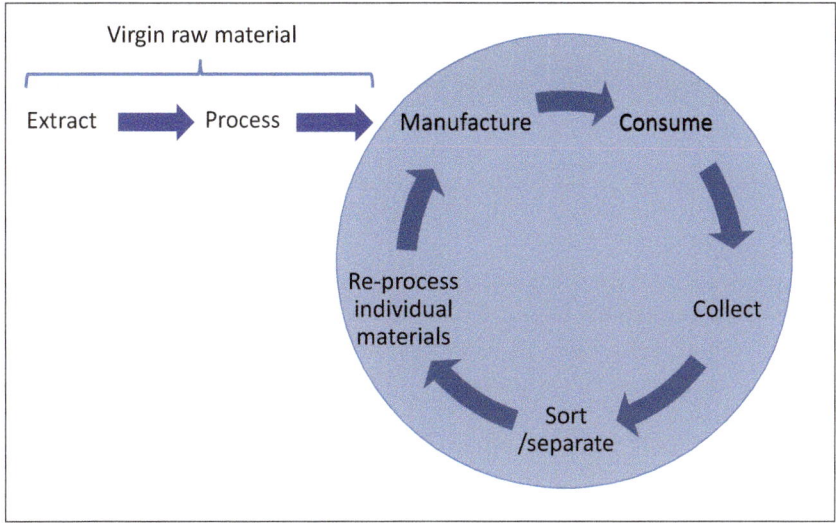

Virgin raw material

Extract → Process → Manufacture → Consume → Collect → Sort /separate → Re-process individual materials →

FIGURE 6.4 The theoretical model of circular consumption.

The second option is what happens to kerbside collected Dry Recyclables now. But can this approach be applied to unwanted products? I'll come back to this in a moment.

I've illustrated the ideal circular model again in Figure 6.4. Following the extraction and refining of raw materials, producers manufacture products that we consume. When we have finished with them, these discarded items should be collected and sorted, either for reuse, repair or for recycling, when the component materials can be separated and reprocessed before re-entering the manufacturing stage as raw materials.

Before going any further, I think it's important to think about the Circular Supply Chain separately for products and for packaging.

Circular treatment of products

Here we're talking about how to manage products that have reached the end of their useful life. The first question to ask is, if they are broken or damaged, can they be repaired by the original owner and thus returned to being used and valued. If so then great, but what if the original owner can't repair the product, can we expand the existing, limited repair industry to bring back the culture of repair?

Or what if the product owner just doesn't want it any more, they're bored with it or having a "clear out"? Well then maybe the product can be reused by someone else giving it a second life. This might be through the original purchaser selling it using a website like eBay, or by donating it to a charity shop.

But if a product can't be repaired or just isn't wanted by anyone, this is where the concept of the Circular Supply Chain should come in. First the unwanted products need to be collected; how can we collect unwanted products?

There is no single answer to this, but rather a series of options. Unwanted products could be:

- forced back down the Supply Chain, as I've suggested, by the retailer who is selling the consumer a replacement product, taking back the unwanted product as some retailers do now, for example for electrical white goods or furniture; but what if the owner of the unwanted product doesn't want a new one, they just want to get rid of the old one; a retailer can't be required to take back an unwanted product if they didn't sell it in the first place and are not selling a replacement, or can they?
- collected via HWRCs (this is my second option above of extending the Supply Chain beyond the consumer); products such as fridges, freezers, televisions, vacuum cleaners and other electrical and electronic products are collected at most HWRCs now, but if a householder can't get their unwanted product to an HWRC, for example if they don't have a car; then
- (for larger items such as furniture) collected by the WCA through their "bulky waste collection" service.

All of the above collection options really only apply to larger unwanted products. What about smaller WEEE products such as phones, electric clocks, radios, cameras, memory sticks, or an old computer mouse. All of these contain very valuable, but hard to extract, materials. Some, but few, WCAs collect these as part of their kerbside collection (but there are doubts about the safety of this in co-mingled collection (see Chapter 10)). Many HWRCs have enormous skips for people to tip all smaller WEEE items into (but I have to say I'm dubious about what happens to this mixture of WEEE). We therefore need a method of collecting smaller products, prior to dismantling and reprocessing.

And what about other products such as clothes, household kitchen goods or small household furnishing items such as table lamps?

I think the collection of small products requires a partial reversal of the Supply Chain. Some retailers have started to collect unwanted products that are relevant to their business, which they then feed into the dismantling/reprocessing industries. For example, I know of one large DIY retailer that collects unwanted small WEEE items, light bulbs and batteries and one large grocery retailer that allows customers to deposit unwanted household batteries and water filter cartridges. But these are isolated examples. We need

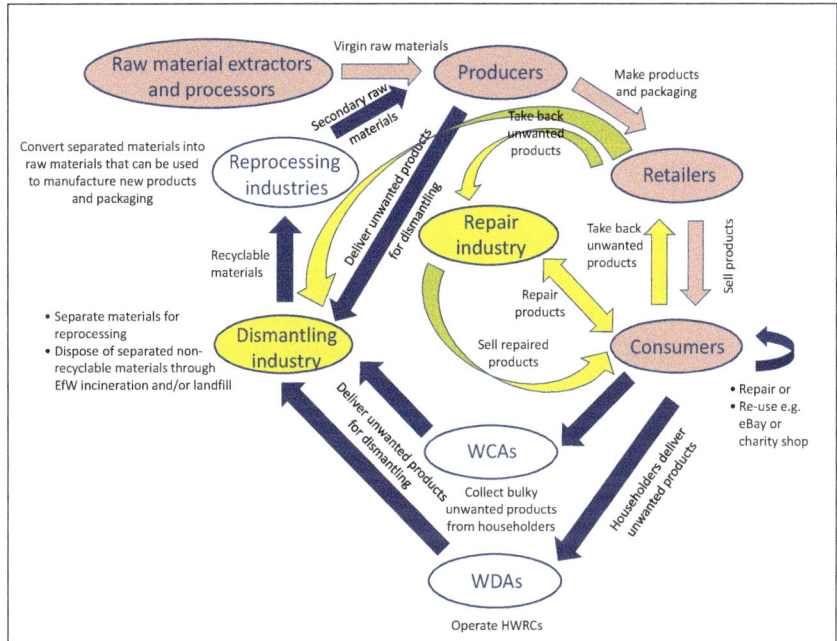

FIGURE 6.5   Multi-channel collection of unwanted products.

a much more organised approach whereby all large retailers allow consumers to bring in unwanted products that the retailer collects and uses the fleets of delivery trucks to transport these collected products either for repair or dismantling.

This multi-channel collection option is illustrated in Figure 6.5, which shows:

- the current linear Supply Chain in pink;
- the existing collection routes for no longer wanted products in blue; and
- the new, required collection infrastructure in yellow.

As Figure 6.5 shows, we need three major changes to occur, if we are to create anything like a Circular Supply Chain for products:

- first, we need to establish two new industries:
  - a repair industry comprising specialist repairers who will take those products that are suitable for repair (in terms of being designed for repair and being of an acceptable quality); and
  - a dismantling industry that takes in unwanted products to separate them into individual materials that can be sold to reprocessors for conversion into secondary raw materials;

- second, we need to create the collection infrastructure shown in Figure 6.5, comprising:

  - the existing collection methods using HWRCs and WCA bulky "waste" collection for large products; and
  - a new collection route for small products whereby retailers take back unwanted products that are relevant to their business, to send these on for repair or dismantling; and

- we need an attitudinal change by consumers, whereby they use the existing and new collection methods to feed unwanted products into the repair/dismantling infrastructure (this will be a major, major attitudinal shift from the current attitude of the majority of consumers of just wanting to discard what they no longer want).

I've already talked about how an expanded repair industry could be created, but having collected unrepairable products, who will dismantle them and how?

This is about harvesting the valuable materials that were used to make the product, by separating and then reprocessing them to make new products.

The "who" should surely be the industry that produced the product in the first place. I am not suggesting individual producers take back their own products for dismantling, but that, for example, the producers of electrical white goods (e.g. fridges, freezers and washing machines) set up dismantling operations to separate the valuable materials contained in the unwanted products; similarly with mobile phones. This could lead to the creation of a whole new industry dedicated to dismantling unwanted products. I can foresee a cascading network of dismantlers, starting with the dismantling of major products, for example white goods into component parts, such as electric motors, electronic circuit boards, electrical switches and controllers, and then each of these component groups being sent to specialist dismantlers of the specific components.

And this leads to the obvious question of who is going to pay for these dismantling and separation activities. The answer has to be the consumer. End-of-life treatment has to be a part of the cost of consumption. It will be anyway, as if producers are required to up their game through Extended Producer Responsibility, they are going to pass on the costs of this to the consumer and I think rightly so.

But the big question for me is, assuming retailers can be persuaded to take back smaller products and assuming an expanded repair industry and a new dismantling industry can be created, will consumers play their part in delivering unwanted products for treatment. This is the key question for me when we talk about a Circular Supply Chain for products.

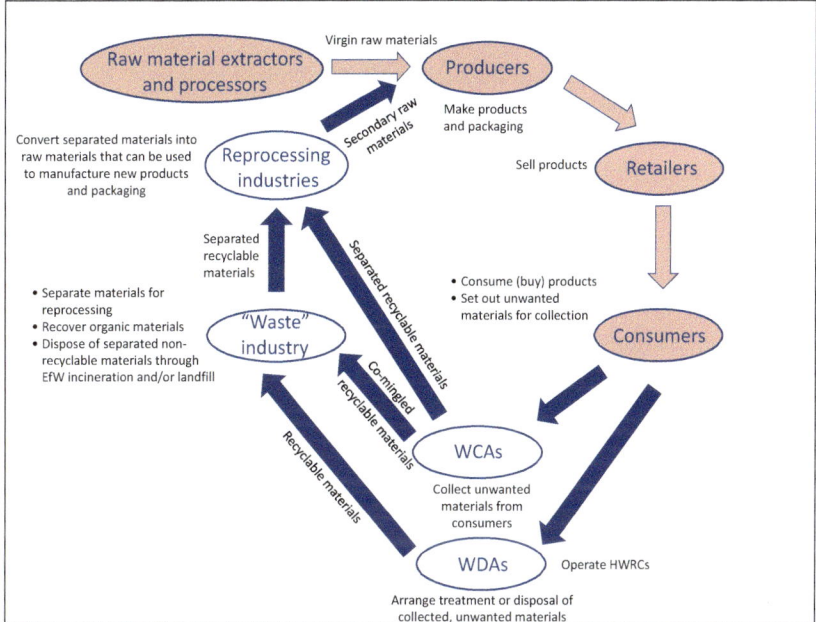

FIGURE 6.6   The Circular Supply Chain for packaging.

Circular treatment of packaging

I think creating a Circular Supply Chain for packaging materials is much more straightforward than for products, for two reasons.

First, if we can implement effective kerbside collection arrangements for all households, then we have addressed the collection problems identified above for products.

Second, most packaging comprises single, separated materials that can be sorted in a MRF. If more than one material is combined, either the packaging needs to be redesigned (remember my example of yoghurt pots earlier) or reprocessing approaches will have to be developed to separate the materials as part of the reprocessing stage of recycling (for example Tetra Pak type cartons), or this type of packaging has to be phased out (although I can't really see this happening).

The Circular Supply Chain for packaging is illustrated in Figure 6.6 and you will see that this Circular Supply Chain has no yellow elements; all of the infrastructure exists today to make this happen.

But not all packaging is suited to a fully Circular Supply Chain.

Paper and cardboard packaging are a good example of materials that do not fully fit the circular model. Yes, paper and card can be collected for recycling and reprocessing in paper mills but, as I've already said, whenever paper and card are reprocessed, a proportion of the fibres in the paper and card break and become too short to form part of recycled paper or cardboard; these fibres are thus discarded within the reprocessing stage of the cycle. Paper and cardboard can only be recycled five to seven times before the fibres become too stiff and too short to make new paper.[3] But also, not all paper products are suitable for recycling, for example used toilet tissue, paper tissues and cardboard products that have become greasy (such as pizza boxes and fish and chip wrappers) cannot be reprocessed. One figure I have seen suggests that only 80% of paper products are suitable for recycling.[4]

Or take glass bottles and jars as another example. Some glass containers are broken before they can be collected for recycling and so are excluded from kerbside collection schemes to protect the kerbside operatives. Whilst the process to recycle glass is very efficient, not all glass packaging ever reaches a reprocessing plant. The same applies to metal cans, both steel and aluminium, many of which are not collected for recycling because they are discarded in mixed litter bins, or even worse, chucked away as litter.

The above examples demonstrate that there is leakage from the theoretical Circular Supply Chain and it is important that we recognise this. We can't expect to have a 100% Circular Supply Chain with everything produced the first time round necessarily having a second, third, fourth or infinite life. Obviously, we want to minimise this leakage, but in nearly all cases it can't be eliminated.

This leads me to three conclusions:

- the efficient and effective **collection** of unwanted products and packaging is critical to the success of a Circular Supply Chain, but is not being given sufficient attention by those organisations and producers currently developing so-called circular solutions;
- in the case of many materials, the Circular Supply Chain requires topping up every time we go around the cycle, due to leakage; and
- the materials that "leak" from the Circular Supply Chain still need to be captured and dealt with, in as an environmentally sustainable way as possible, for example by combustible materials such as paper reprocessing rejects being sent for EfW incineration.

This more realistic circular model is shown in Figure 6.7, which shows that some of the consumed materials:

- will leak either as a result of:
    - our consumption, e.g. consumers just not setting out unwanted materials for recycling, for example by putting packaging into their

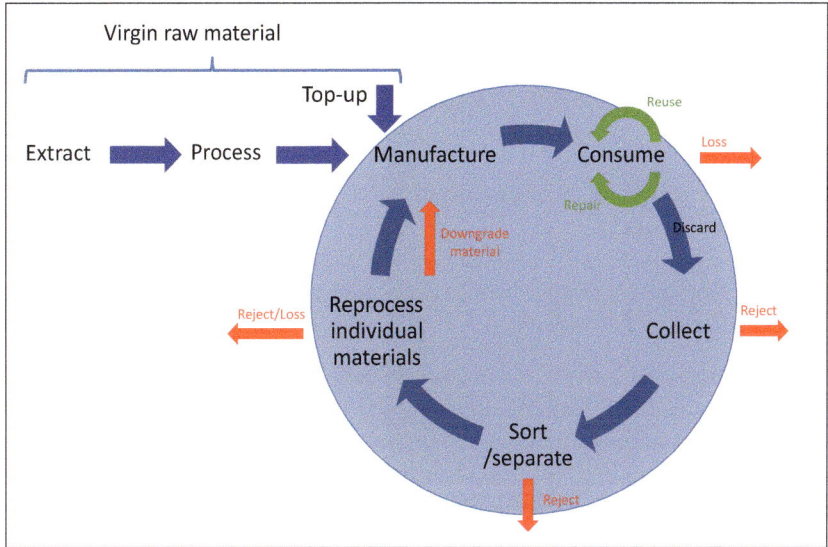

FIGURE 6.7 A more realistic model of circular consumption.

kerbside bin for disposal or public litter bins, or worse just littering unwanted packaging on the ground; or

- losses during reprocessing, e.g. paper reprocessing losses;

- are rejected during reprocessing when incorrect materials are identified and discarded, e.g. during kerbside sorting, at a MRF or during reprocessing; or
- are down-graded during reprocessing as not all materials can be converted back into virgin-equivalent grade materials for remanufacture (down-cycling).

Such losses/rejections have to be made up for by topping up with an input of new virgin raw material to the Circular Supply Chain. So, such a supply chain is not truly circular and I think it would be idealistic, if not naïve, to think it ever could be. But as an ideal, it is an excellent target to aim for.

Conclusions

To create Circular Supply Chains for products we need:

- to reuse or repair products to give them a second, third or fourth life, either with the original owner or passing them onto someone else who wants them;

- to put in place comprehensive arrangements to collect unwanted products, via a combination of:
  - retailers taking back an old, unwanted product every time a consumer buys a new product and retailers facilitating the deposit by consumers of unwanted smaller products in their stores and then sending these either for repair or dismantling by a partial reversal of the current Supply Chain;
  - bulky products to be collected at HWRCs as now;
  - WCAs to collect larger products from households through their "bulky waste collection" service;
- the **industry** (not the original producer) that produced the original product to be responsible for setting up both repair and dismantling operations to separate the valuable materials contained in the unwanted products;
- to transform the attitude of all consumers so that they use the enhanced collection systems for unwanted products; and
- the end-of-life treatment of products to be a part of the cost of consumption paid for by the consumer.

To create Circular Supply Chains for packaging we need to:

- implement effective kerbside collection of Dry Recyclables from all households;
- develop reprocessing approaches to deal with packaging in which more than one material is combined, or phase out this type of packaging all together;
- accept that not all packaging is suited to a fully Circular Supply Chain, but give the materials concerned as many lives as possible; and
- recognise that for many materials, the Circular Supply Chain requires topping up every time we go around the cycle, due to leakage.

But before talking about what we need to do differently, let me take a step back and describe briefly, why we consume so much.

Notes

1 David Attenborough, *A Life on Our Planet*, Witness Books, 2020, p. 204.
2 For example, see Peter Lacy & Jakob Rutqvist, *Waste to Wealth*, Palgrave Macmillan, 2015, pp. 4–5.
3 The New York Times, "Through the Mill", 20 December 2010, www.nytimes.com, accessed 6 November 2020.
4 https://thecpi.org.uk/library/PDF/Public/Publications/Other/Myths%20and%20 the%20Facts.pdf, 2 February 2017, accessed 6 November 2020.

7

CONSUMERISM AND ECONOMIC GROWTH

I'll start with a quotation from Professor Tim Jackson, who wrote a book called *Prosperity Without Growth* to which I will return later:

> So why is it that commodities continue to be so important to us, long past the point at which material needs are met? Are we really natural born shoppers?[1]

The answer to this question I think is "yes", but that doesn't mean we can't change. There are at least three reasons why I think we are currently "natural born shoppers":

- human evolution and hard-wiring in our brains;
- social pressure and our creation of the consumer society; and
- economics, capitalism and our obsession with economic growth.

I'll take each of these in turn, but the key thing here is, as Madonna sang in 1984, "We're living in a material world …"; sadly, this is so true. But, for those of us fortunate to live in developed countries, for many of us, we already have enough stuff, so why do we continue to desire and acquire more material objects?

Human evolution and hard-wiring in our brains

According to Professor Jackson "Our brains evolved in an ancestral environment of relative scarcity … Our propensity to over-consume 'is a relic

DOI: 10.4324/9781003504757-8

of a time when individual survival depended on fierce competition for scarce resources' ".

"A core element of our neural design is that a pulse of dopamine is delivered to key areas of the brain whenever we obtain what … we desire the most". In other words, we feel good about a purchase. However, Professor Jackson goes on to say, "The pulse soon fades, so to obtain another, we repeat the behaviour".

"This relentless repetition is reinforced by a couple of additional design features in the human brain. One of these is … habit … our brains delegate as many decisions as possible to the realm of automaticity".

Thus, according to Professor Jackson, our brains are neurologically hard-wired to consume.

But, whilst I am sure Professor Jackson is correct in explaining the antecedents of our behaviour, does this mean we cannot change? Isn't this a bit of a cop-out? It's like saying "I didn't want to buy this thing I don't really want, but my pre-programmed brain made me do it"?

Surely, as we come to realise the damaging effects of unnecessary and inappropriate consumption and over-consumption, we can ignore the potential dopamine hit and use our cognitive abilities to rationalise what and how we buy things. Isn't it the case that we need to think a little more when preparing to consume something and rescue our decisions from "the realm of automaticity"?

But there is more.

Social pressure and our creation of the consumer society

We are constantly bombarded with advertisements telling us how we could have a more fulfilled and satisfied life if only we bought the new Product X or Product Y. Whether this advertising is on television, on social media or in print in magazines or newspapers, or just walking down a high street, we are constantly urged to buy more stuff. And don't get me started on the subject of so-called social media influencers. The underlying message is that if you don't buy more stuff, you are somehow failing to fully participate in modern life.

You haven't replaced your kitchen in the last five years or your bathroom? Shame on you. You don't have a kitchen tap that instantly dispenses boiling hot or chilled water? What are you thinking?

And the most pernicious aspect of this marketing and advertising is that it can make people feel inadequate if they think they aren't keeping up with everyone else. To again quote Tim Jackson, "In mid-2008, shortly before the collapse of Lehman Brothers … Households in the UK … were 'maxing out their credit cards' and running down their savings just to stay in the game … a story of ordinary people spending money they don't have, on things they don't need, to create impressions that won't last on people they don't care about".

What a damning, but sadly true, description of our current consumer society. Just take a moment and read that quotation again.

So, the next time you are tempted to buy something, just ask yourself, do I really need this or do I just want it? And if the latter, why do I want it? Is it because it will significantly improve my life or because everyone else is buying one? Just because you can (buy something) doesn't mean you should.

But there is even more.

Economics, capitalism and our obsession with economic growth

Every time there is a national financial crisis, for example as occurred in the UK in 2022, the Government's knee-jerk reaction is to make changes to drive up economic growth. The then Prime Minister, Liz Truss, announced a disastrous mini-budget in September 2022 which was all about "growth, growth, growth". The mantra is, if we increase economic growth then all will be well.

And for many years this has proved to be the case. Why? Well let's take a look at what economists call "the engine of growth", which is illustrated in Figure 7.1.

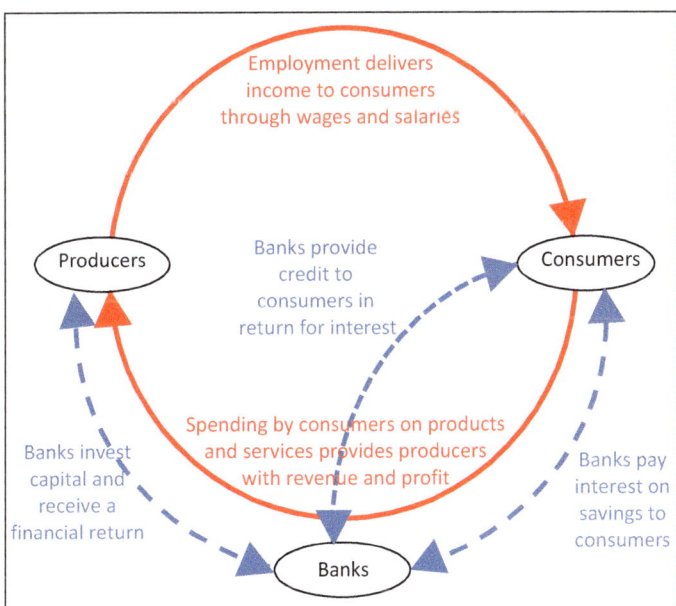

FIGURE 7.1   The economic engine of growth.

Starting at the left-hand side of Figure 7.1 and following the red circle first, producers make products and deliver services and employ people to make this happen. These employees receive wages and salaries, which allow them as consumers to buy products and services, thereby closing the circle and allowing producers to earn income and make a profit and so produce more products and services.

This simple circular model becomes a little more complicated when we add in the concepts of credit and savings (the blue dotted lines). Credit is when banks and other lending institutions allow consumers to spend money they don't have, in return for (banks) receiving interest payments. This allows consumers to live beyond their immediate means and risks them building up debt, which will one day have to be repaid.

On the other hand, when consumers have more money than they need for their day-to-day consumption, they can choose to save their excess income with a bank, and in return they receive interest on their savings.

With the money that banks hold on behalf of savers and the profits they themselves earn, they invest in producers for a financial return. This investment helps producers to grow, expanding their operations, employing more people, who can afford to buy more products; and so the circle continues.

So why does this circular engine of growth need to grow? Why do producers have to employ more people to produce more products and services? The answer is simple. It is because the human population is continually growing. In 2011 the world population was seven billion, on 15 November 2023 it was expected to hit eight billion and in 2037, the prediction is nine billion, according to the UN.[2]

What I have described above and illustrated in Figure 7.1 is a gross over-simplification and I am no economist, but it serves my purpose for now. My point is that we need economic growth just to keep up with population growth, never mind helping poorer nations catch up with the affluent developed world. But as ever, there is more to it than this.

Going back to Tim Jackson, he says "The modern economy is structurally reliant on economic growth for its stability. When growth falters … politicians panic. Businesses struggle to survive. People lose their jobs and sometimes their homes".[3] We can see from Figure 7.1 how the economy can quickly turn into a downward spiral.

The role of capitalism in driving economic growth

Now we come to a really interesting and sensitive subject. As I said I am no economist, so I will defer to others to explain capitalism and why it creates a fundamental problem for human kind.

I'll start with a quotation from Jason Hickel, a well-known writer on economic inequality:

> What's ultimately at stake is the economic system that has come to dominate more or less the entire planet over the past few centuries: capitalism.[4]
>
> We have a tendency to describe capitalism with familiar, well-worn words like "markets" and "trade". But this isn't quite accurate. Markets and trade were around for thousands of years before capitalism, and they were innocent enough on their own. What makes capitalism different to most other economic systems in history is that it's organised around the imperative of constant expansion or "growth": ever increasing levels of industrial production and consumption, which we have come to measure in terms of Gross Domestic Product (GDP). Growth is the prime directive of capitalism. And as far as capital is concerned, the purpose of increasing production is not primarily to meet specific human needs, or to improve social outcomes. Rather, the purpose is to extract and accumulate an ever-rising level of profit. This is the overriding objective.

He goes on to say:

> Under capitalism, global GDP needs to keep growing by at least 2% to 3% each year, which is the minimum necessary for large firms to maintain rising aggregate profits. ... Three percent growth means doubling the size of the global economy every twenty-three years, and doubling it again ... and again.
>
> As production increases, the global economy churns through more energy, resources and waste each year, to the point where it is now dramatically over-shooting what scientists have defined as safe planetary boundaries, with devastating consequences for the living world.[5]

I'll come back to safe planetary boundaries in a moment.

So, I have cited above two eminent writers who are eloquently making the case that our economic model of capitalism is based on the presumption of continuing growth.

But, continuing economic growth as we know it cannot continue indefinitely, for one reason and one reason alone. Economic growth is predicated on consuming more and more of the Earth's finite resources to fuel that growth. And as Tim Jackson says, "We're running out of planet".[6]

But as Tim Jackson goes on to say, "The idea on a non-growing economy may be anathema to an economist. But the idea of a continually growing economy is anathema to an ecologist".[7]

So, what are we to do?

De-growth?

As more and more people have come to realise that we cannot maintain our obsession with economic growth, a movement has emerged which advocates what is called "de-growth". Jason Hickel describes de-growth as "… a planned reduction of excess energy and resource use to bring the economy back into balance with the living world …"[8]

This is where we start to enter sensitive areas. On the one hand, the governments of developed countries will say that they need continuing economic growth, in order for their citizens to continue to thrive, whilst those of developing countries will argue they need economic growth so their citizens can catch up with the developed world and share in the benefits this would bring.

Governments seem to want economic growth to continue indefinitely. But how can it, when our planet is finite and in many cases is already being over-exploited?

Planetary boundaries

I said earlier that I'd come back to the idea of safe planetary boundaries, so here goes.

Again, quoting Tim Jackson, he said

In 2015, the Stockholm Resilience Centre published its second "planetary boundaries" report … [which] … carried out an extensive audit of our proximity to nine "critical biophysical boundaries". Crossing these boundaries … would imply unacceptable environmental change with "serious, potentially disastrous consequences" for society. … The team discovered that current levels of economic activity already lie beyond the "safe operating space" of the planet, for four of these [nine] critical boundaries.[9]

This idea of the safe operating space of the planet and the nine critical boundaries has been developed in a powerful way by the economist Kate Raworth in her book "Doughnut Economics".[10] I highly recommend her TED talk on Doughnut Economics.[11]

The nine critical boundaries are:

- climate change;
- ocean acidification;
- chemical pollution;
- nitrogen and phosphorus loading;
- fresh water withdrawals;

- land conversion;
- biodiversity loss;
- air pollution; and
- ozone layer depletion.

Nine very critical boundaries, but where is natural resource depletion? Surely, running out of planet should be included here? I am amazed that it is not.

Doughnut economics

What Kate Raworth did was to take the nine critical planetary boundaries and combine them with where our global economy is currently falling short in delivering what every human needs. The planetary boundaries form an outer circle, which we should not go outside. The shortfalls form an inner circle, which we should not go within. Between these two circles is where our global economies should be operating. This is the doughnut. She says:

> So imagine humanity's resource use radiating out from the middle. That hole in the middle is a place where people are falling short on life's essentials. They don't have the food, health care, education, political voice, housing that every person needs for a life of dignity and opportunity. We want to get everybody out of the hole, over the social foundation and into that green doughnut itself.
>
> But, and it's a big but, we cannot let our collective resource use overshoot that outer circle, the ecological ceiling, because there we put so much pressure on this extraordinary planet that we begin to kick it out of kilter. We cause climate breakdown, we acidify the oceans, a hole in the ozone layer, pushing ourselves beyond the planetary boundaries of the life-supporting systems that have for the last 11,000 years made Earth such a benevolent home to humanity.
>
> So, this double-sided challenge to meet the needs of all within the means of the planet, invites a new shape of progress, no longer this ever-rising line of growth, but a sweet spot for humanity, thriving in dynamic balance between the foundation and the ceiling.

And she is quite clear and damning about where we are today: "Look in that hole, you can see that millions or billions of people worldwide still fall short on their most basic of needs. And yet, we've already overshot at least four of these planetary boundaries, risking irreversible impact of climate breakdown and ecosystem collapse. This is the state of humanity and our planetary home".

So, what to do?

The future of global economics

I've already said that I am no economist, so I am far from qualified to suggest the solution to our current economic problems and failures. I have quoted three eminent economists who are thinking about these issues and they have all in different ways come to one common conclusion.

Tim Jackson says "… the idea of a continually growing economy is anathema to an ecologist". Jason Hickel advocates "… a planned reduction of excess energy and resource use to bring the economy back into balance with the living world". And Kate Raworth tells us we have to move billions of people out of the hole in the centre of the economic doughnut, without overshooting the nine critical planetary boundaries.

They seem to be in strong agreement that the basic tenet of capitalism, the economic model that has dominated since the 1950s, of unfettered and continuous economic growth, cannot continue on a finite planet. Each in their own way advocates a very significant shift in economic thinking, planning and action to a model that allows people to flourish, rather than our current economic model based on profit and GDP, without thought for people or the planet. I particularly like the idea of allowing people to flourish.

Tim Jackson puts this quite succinctly: "If we forget momentarily about the relentless pursuit of growth, we can concentrate on defining what an economy should look like. Surprisingly it boils down to a few obvious things. Capabilities for flourishing. The means to a livelihood. Participation in society. A degree of security. A sense of belonging. The ability to share in a common endeavour and yet pursue our potential as individual human beings".[12] Who wouldn't want this?

What these thinkers are advocating is a fundamental shift in focus, from one of unrelenting economic growth to providing the foundations that allow people to flourish and always within the finite boundaries of our precious planet. This is what we urgently need, but how this can be achieved on a national and global basis is beyond the scope of this book.

Notes

1 Tim Jackson, "Prosperity without growth", Routledge, 2017, p. 68.
2 Will Pavia, *The Times*, "The world population passes 8 billion today. How big will it get?", 15 November 2022.
3 Tim Jackson, *Prosperity without Growth*, Routledge, 2017, p. 21.
4 Jason Hickel, *Less is More: How Degrowth Will Save the World*, Penguin Books, 2022, p. 18.
5 Jason Hickel, *Less is More*, pp. 18–20.
6 Tim Jackson, *Prosperity without Growth*, Routledge, 2017, p. 17.
7 Tim Jackson, *Prosperity without Growth*, p. 21
8 Jason Hickel, *Less is More: How Degrowth Will Save the World*, Penguin Books, 2022, p. 29.

9 Tim Jackson, *Prosperity without Growth*, Routledge 2017, p. 17. If you are interested, the four critical boundaries were: excess nutrient loading; species loss; ocean acidification; and climate change.
10 Kate Raworth, *Doughnut Economics*, Random House Business, 2017.
11 Kate Raworth, "A healthy economy should be designed to thrive, not grow", TED, youtube.com, April 2018, accessed 8 December 2022.
12 Tim Jackson, *Prosperity without Growth*, Routledge, 2017, p. 219.

8

LET'S TALK ABOUT PACKAGING

Before presenting my detailed proposals for change, I'd like to take a moment to focus on one of the key components of Household Waste, which is packaging. As I said in Chapter 2, packaging makes up about one third of Household Waste. Whilst it is easy to pick on packaging as a problem we need to address, it is actually one of the areas where we are starting to see some progress in its design, collection and recycling. So, let's take a closer look.

Why packaging is needed

Packaging protects products from damage during handling and transport, facilitates the display of the product and, in the case of food, can both prolong the life of the product and provide anti-tampering protection.

Let's look at one example that shows the benefits of packaging, the shrink-wrapping of cucumbers. In their book *Why Shrink Wrap a Cucumber?*[1] Lauren Miller and Stephen Aldridge say "… a wrapped cucumber lasts more than three times as long as an unwrapped one …" Some might argue that given all the cost and time that has gone into producing a cucumber, surely wrapping it in shrink-wrap plastic is worthwhile to protect it and prolong its useful life? So it would appear; however, the counter-argument starts with the fact that locally produced and sold cucumbers are not shrink-wrapped. They are moved from farm to retailer to consumer without the protection of plastic film and once stored in a fridge last as long as a wrapped one.

It would appear that the real reason why the protective packaging is needed is because products like cucumbers in our supermarkets are produced on an industrial scale and transported long distances and so need

DOI: 10.4324/9781003504757-9

the protection and life-enhancing properties of plastic packaging, in order for this large-scale production, distribution and retailing to function. Is it not the case that if food were grown and retailed locally then the benefits that such packaging provides would not be needed? Isn't the answer to such packaging problems (and shrink-wrapping plastics are not recyclable) to source and buy locally?

But I think there is a much deeper issue here, which is all to do with consumers' desire for convenience and ease of shopping. The whole reason that supermarkets have developed is the concept of the (quite literally) one-stop shop. A grocery supermarket offers just about everything that consumers want in terms of food and drink, all under one roof. It's really convenient for the consumer, but products are sourced and distributed nationally (some are sourced internationally), squeezing out local supply (because the supermarket chain wants a consistent offering across all of its stores and often across all seasons). This necessitates high levels of protective packaging.

Packaging also often allows producers to promote their product, through attractive branding and design and the inclusion of information about the product, both promotional and statutory.

But if we accept that packaging plays vital and useful roles, why is it the focus of so much negative attention? Part of the reason is because it is transient. Once a product has been bought and taken home by the consumer, the packaging becomes redundant and is discarded. This is a classic, linear use of resources, unless the packaging is reuseable or recyclable, which in the vast majority of cases it isn't and, if we're honest, never will be.

Different levels of packaging

There are three levels of packaging: primary, secondary and tertiary. Primary packaging is what we as consumers see when we buy a product. Secondary packaging holds the primary packaged products together for convenient handling in-store (I'm sure you've seen cardboard trays holding cans on a shelf, but if you look up to the top of the display shelving, you'll see these same trays of tins wrapped in single-use plastic, waiting to be moved to the display shelf). Tertiary packaging is that used to hold multiple packs of secondary packaging together for transport, often on pallets. I find it appalling that these pallets of products are wrapped in non-recyclable, single-use, plastic shrink-wrap film.

Whilst I understand the need for these three levels of packaging, I am concerned that distributors and retailers focus their efforts on packaging reduction on secondary and tertiary packaging and not on the primary packaging which is taken away by consumers.

Secondary and tertiary packaging is much easier to collect and recycle within the Supply Chain, and so it is understandable that distributors and retailers will focus on this first as it's an easier win. But producers and retailers must also address the issue of primary packaging to minimise it and to make it easily recyclable.

I am also concerned that when retailers make claims that they are reducing the amount of packaging within their supply chain, what they are actually referring to relates to reductions in secondary and tertiary packaging only (have you seen significant changes in the way primary packaging is being used – I haven't). Whilst such reduction efforts are to be applauded, these statements can be misleading and verge on greenwashing.

Reuseable packaging

How often have you seen a statement on a product's packaging that states "This packaging is reuseable" together with instructions on how this can be done? Never? Me neither.

But reuse is right at the top of the Three Rs, so why don't producers and retailers promote it much more? In part because it's actually the retailers, not the producers who would facilitate reuse. We need retailers to put the infrastructure in place to allow packaging reuse to happen, but they will also need support from producers to make this a reality. See Chapter 10 for more discussion on this.

Recycled content

I'm also sceptical when a producer makes a statement such as "This product contains 50% recycled content". Where has the "recycled content" come from? Producers have always recycled materials that are discarded in the manufacturing process, such as excess material that is trimmed off a product or packaging, as this makes good financial sense, particularly as the material is clean, is a single material and is already located within the factory. But is the producer doing more than this internal recycling?

Any "recycled content" claim could be another example of greenwashing, unless the source of the "recycled content" comes from post-consumer sources (as this is the only way that circular consumption can actually happen). So, whenever a producer makes a claim that their product or packaging contains "recycled content", I would like to see the statement including the source of this material, so for example saying, "Contains 50% **post-consumer** recycled content".

Excess packaging

I am sure you have seen many examples of excess packaging, where even more materials are destined for rejection. Examples of such excesses range from plastic-wrapped products such as "handy-size" tissue packs being wrapped in further plastic film to create a multipack. Another example is plastic shrink-wrapped cans of food, such as baked beans or boxes of tissues, again to create multipacks. And what about luxury goods with fancy packaging to tempt us into buying, but which is of no use and is instantly discarded once the product is removed from the packaging?

But there are some producers, for example one brand of tinned tomatoes, that create multipacks using a recyclable cardboard sleeve instead of plastic shrink-wrap. And Tesco recently announced[2] that it would no longer stock cans of beer multipacked using six-pack plastic rings.[3] This begs the very obvious question: if a small handful of producers and retailers can eliminate this excess packaging, why can't all producers do the same?

Inappropriate packaging

As well as excess packaging, there is a phenomenon of using inappropriate materials for packaging. One example that really frustrates me is toilet tissue and kitchen paper towels, the vast majority of which are wrapped in single-use plastic film (again, often to create multipacks). I've managed to find a brand of kitchen towels (but only one) wrapped in paper in my local supermarket and so bought it, simply because of its paper packaging. And now for the first time I have found two brands of toilet tissue that wrap their products in paper – hooray!

There is no reason why all toilet tissue and kitchen paper cannot be wrapped in paper, indeed I seem to remember it used to be, so producers, please stop flooding us with single-use plastic film. And again, if two producers can eliminate single-use plastic packaging in favour of recyclable paper, why can't all producers do the same?

Another example of inappropriate packaging that frustrates me is plastic parcel tape. This is used everywhere, by producers, retailers and by individuals. It's plastic and so all the arguments against plastic use apply, but what makes its use worse is that it is often used excessively. One of the worst offenders for this I have come across is people who send items bought on eBay. They seem to have an almost obsessive zeal to cover every inch of their cardboard packaging with plastic parcel tape. I've shown two examples I have received in Figure 8.1 (I buy a few second-hand things on eBay as part of my drive for reuse, rather than purchasing new).

There are perfectly good and effective self-adhesive paper parcel tapes available, which can be left on the cardboard box to which they have been attached, so when the boxes are put out for recycling the paper tape is recycled as well. So, let's switch from plastic parcel tape to paper (eBayers please take note!) But, as ever, beware. Some paper parcel tape contains plastic filaments to add strength to the tape. These filaments will be treated as contaminants

FIGURE 8.1 Excessive use of plastic parcel tape.

in cardboard reprocessing and so will be sent for incineration or landfill. Are they really needed? If other paper tapes don't use them, why not all such tape?

But things are starting to change – or are they? Miller and Aldridge say "We are at the start of adopting a globally sustainable approach to living. Much is still to be done, but at least we recognize many of the issues. Advances may come with new technology, materials, greener retailing, developments in recycling, legislation or even consumer attitudes".[4]

I wish it were so, but I disagree. We are light years from "adopting a globally sustainable approach to living", and I am convinced that the majority of consumers and most producers and retailers do not "recognise many of the issues". And whilst to say "advances may come with new technology etc." shows a touching faith in the future, it is really just putting off addressing the problem. We need to act now and use the tools and approaches available to us today. The issue is really one of changing attitudes (producers, retailers, WCAs and consumers) and rethinking how we manage the Earth's limited resources. The solutions to our "waste" problem exist in our hands today.

But there is a chink of light. Miller and Aldridge also say, "However, retailers and manufacturers have not only acknowledged the need to address the effect of packaging on the planet, they have also realized that there is a great deal of money to be made by being seen to be green ..."[5] Consumers and the media have made producers and retailers start to realise that they have to address consumers' environmental concerns. But these are very early days and my fear is that there is a lot of greenwashing going on and not so much concrete action.

Packaging labelling

Let's now talk a little more about packaging labelling. As well as the plastic materials identifiers shown in Table 9.1 (see Chapter 9), other symbols and labels appear on packaging to inform us and to encourage us to recycle them. The most common are shown in Table 8.1.

TABLE 8.1 Packaging labelling symbols and their meaning

Symbol	Name	Meaning	Whether packaging is currently recycled
	Mobius loop	The triangular three green arrows simply mean the packaging is capable of being recycled.	Maybe, maybe not, the symbol doesn't tell us this.
	The Green Dot	The Green Dot is a symbol that was first introduced in Germany and is now used on packaging in a number of European countries to signify that a producer has made a financial contribution to a national packaging management scheme, financing the sorting, recycling or recovery of their packaging when it eventually becomes "waste"* This symbol has no meaning in the UK and usually appears on products here that are sold across Europe.	The Green Dot symbol simply indicates that the producer has joined the Green Dot scheme and does not necessarily indicate that the packaging on the product or the product itself is recyclable or is recycled.
compostable	Seedling symbol	The product is certified to be industrially compostable according to the European standard EN 13432. When successfully certified, the product will fully biodegrade in an industrial composting plant under controlled conditions such as temperature, moisture and time frame, leaving nothing behind but water, biomass and CO_2.** But this doesn't mean that the product will necessarily decompose in a domestic compost heap.	This depends on whether the WCA accepts the packaging in its food or garden waste collection and, remember, so-called compostables may be rejected from anaerobic digestion and composting plants as contaminants and compostable plastics are a contaminant in plastics recycling (see Chapter 9: Biodegradable plastics and why they are a bad idea).

(Continued)

TABLE 8.1 (Continued)

Symbol	Name	Meaning	Whether packaging is currently recycled
FSC Certified fibre	FSC Certified fibre	100% of the fibre in the product comes from a Forest Stewardship Council (FSC) certified forest, that is managed with consideration for people, wildlife and the environment.	Positive for the environment, but not relevant to recycling.
FSC Certified Recycled	FSC Certified Recycled	All fibre in the product must be pre- or post-consumer reclaimed. For wood products, a minimum of 70% fibre must be post-consumer reclaimed. There is no threshold for paper products, but all inputs must be verified as reclaimed.	Shows a high proportion of the wood fibre in the product is from recycled sources. But not whether the product is recyclable.

Notes:
* Taken from the complydirect.com website, 9 June 2010.
** https://docs.european-ioplastics.org/2016/publications/EUBP_Guidelines_Seedling_logo.pdf, accessed 18 May 2021.

But there is another group of symbols that really winds me up. These are called On-Pack Recycling Labels (OPRLs) and I've shown some examples in Figure 8.2. These are the labels that producers include on packaging to supposedly tell us whether the packaging is recyclable or not.

The OPRLs illustrated in Figure 8.2 are at best misleading and at worst untrue. They are really examples of producers not taking responsibility for their products or packaging at the end of their useful lives.

Why do I say these statements are unacceptable? Take the first statement "TRAY Check Locally". What does this actually mean? I think it means that the tray is in theory recyclable (this statement usually refers to plastic trays, not cardboard), but the producer doesn't know if your WCA/WDA actually accepts such items either in the kerbside collection or at HWRCs. It's fair

FIGURE 8.2 Unacceptable packaging labelling.

Note: Reproduced by the author from various packaged products.

enough that the producer doesn't know all the local arrangements; however, this statement is to my mind a cop-out on the part of the producer. If the item of packaging is truly recyclable, then it will be acceptable both in the kerbside collection and at HWRCs. The producer should make the effort to find out if the item is recyclable and make a definitively, clear statement if it is. Don't fudge the issue and give the impression of recyclability, whilst hiding behind a caveat that stops the producer being held accountable (I refer you to my definition of "recyclable" in Chapter 3).

Now the big one, the statement "FILM Not Yet Recycled". This is an absolute abdication of Producer Responsibility (see Chapter 5). Who is the producer assuming will magically introduce the recycling of the product that is "Not yet recycled"? If this material is not currently recycled, let's be honest, it's pretty unlikely that it will ever be, so what the producer is actually saying is "Not recycled". At least the recycling symbol with the diagonal line through it is honest. But the statement "Not yet recycled" is unhelpful and unacceptable. Either the producer needs to campaign to make the labelled item recyclable or they should stop using the material. This is why I say this is the big one;

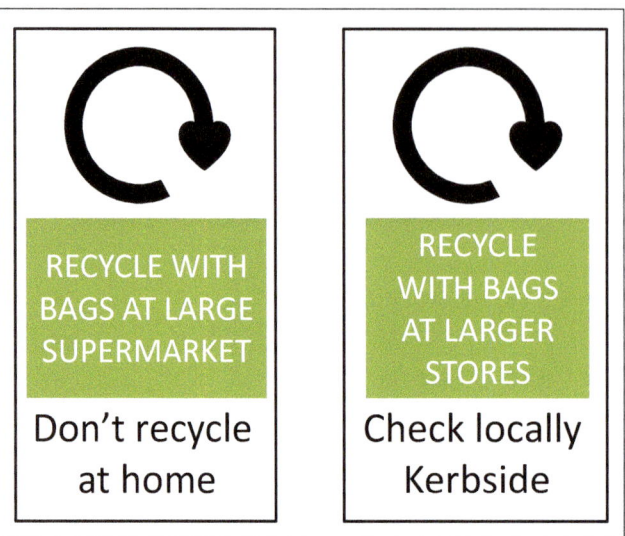

FIGURE 8.3 Examples of producers shifting the responsibility for recycling.

I want to see all packaging materials that are not recyclable phased out and only truly recyclable materials used to package our products.

Figure 8.3 shows two OPRLs that appear on plastic film wrappers and bags and which raise two interesting points. First, the left-hand label is honest when it says "Don't recycle at home", because the producer knows that the WCA/WDA won't take this packaging item (most won't take any plastic film at all); whilst this is honest, it is deeply disappointing. And then the statement "Recycle with bags at large supermarkets" is again a real cop-out by the producer.

Only very large supermarkets have drop-off bins for plastic bags and what they want to collect are unwanted plastic carrier bags made from Low Density Polyethylene (LDPE). They don't want just any old plastic wrapper, particularly if they are not made from LDPE (many product wrappers are made from Polypropylene (PP)). So, the producer is copping-out by expecting two things:

- that consumers will make the effort to collect, store and then take their unwanted plastic film wrappers to a large supermarket (which includes the breathtaking assumption that all consumers shop at large supermarkets); and
- that large supermarkets will both provide and service suitable drop-off collection bins and will accept all plastic films deposited in them and then arrange for the reprocessing of these plastics.

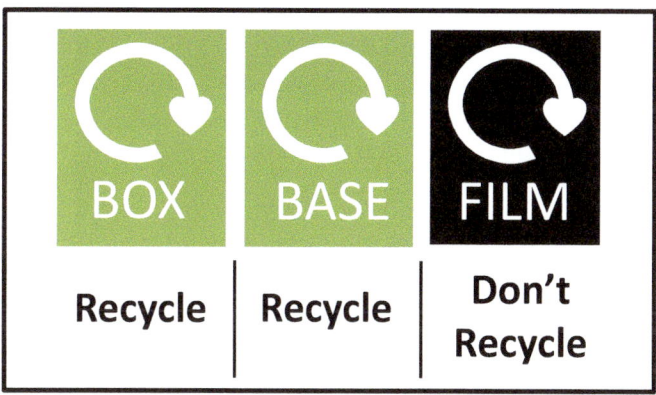

FIGURE 8.4 More honest examples of OPRLs.

Both of these assumptions are seriously flawed and the OPRLs shown in Figure 8.3 are simply the producer greenwashing their product. It is irresponsible and dishonest and should stop.

At least the label shown in Figure 8.4 is honest and states the position clearly when it says "FILM Don't recycle". But clearly, this is not what we want. If a plastic film is going to be used by the producer then it has to be recyclable or it needs to be replaced by something else that is. Simply saying "Don't Recycle" is not acceptable. Producers who use such labels are not taking responsibility for the end-of-life of their products or packaging.

Recycled content

Many producers burnish their green credentials by stating on their products that either the products or packaging are made from materials with a stated proportion of recycled materials.

In April 2002, the UK Government introduced a Plastic Packaging Tax to be applied to all plastic packaging manufactured in or imported into the UK that does not contain at least 30% recycled plastic. The initial level of the tax was £200 per tonne of plastic.

There is a lot of detail contained within this legislation, such as what types of plastic packaging are affected, how the weight of plastic in a mixed material piece of packaging is assessed and whether pre-consumer plastic is included.

Do I think this is this a good idea? To be honest – no, for two reasons:

- by including pre-consumer "waste" plastic within the 30% recycled content, many producers will probably be able to comply now, without making any changes to their use of recycled plastic; it is the demand for post-consumer material that needs to be driven up, so this strikes me as a missed opportunity; and

- it is possible that some producers (or importers) will simply pay the tax and pass it onto their customers, thereby making no difference to the demand for recycled material.

Packaging design

As I have already said, we need producers and packaging designers to design-with-the-end-in-mind.

Obviously we want packaging to be minimised, whilst still performing its functions of:

- protection in distribution and the retail store;
- anti-tampering;
- making the product look attractive to encourage consumers to buy;
- providing statutory information about the product; and
- providing marketing messages.

But there are some very simple principles that we need designers to adopt:

- consider whether the packaging could be made to be reuseable;
- only use packaging materials that meet my definition of recyclable; and
- if you must use multiple materials in one piece of packaging, make them easy to separate by the householder.

There is an urgent need to educate product and packaging designers to adopt these principles and particularly to design-with-the-end-in-mind and we could do worse than starting with the education and training that new designers receive as part of becoming designers.

Conclusions on packaging

Packaging is a necessary evil in our consumer society, so we need to find ways to make it reuseable or recyclable.

Producers and retailers are already making changes to do this for secondary and tertiary packaging, so we need them now to focus on facilitating primary packaging solutions.

Producers need to stop greenwashing in their claims about:

- the recycled content of their packaging; and
- the recyclability of their packaging;

And it needs to be made clearer by producers and retailers what can and what cannot be recycled. So, we need an overhaul of the OPRLs that are printed by producers on their packaging.

Notes

1   Stephen Aldridge & Lauren Miller, *Why Shrink Wrap a Cucumber?*, Lawrence King Publishing Ltd, 2012, p. 42.
2   The *Sunday Times* online newspaper, "Tesco scraps six-pack plastic", 9 May 2021.
3   Such plastic packaging is highly dangerous to wildlife such as seagulls (particularly at landfill sites) that get their heads or feet caught in the loops of the multipack banding.
4   Stephen Aldridge & Lauren Miller, *Why Shrink Wrap a Cucumber?*, Lawrence King Publishing Ltd, 2012, p. 28.
5   Stephen Aldridge & Lauren Miller, *Why Shrink Wrap a Cucumber?*, p. 27.

9
A FOCUS ON PLASTICS

An introduction to plastics

Now let's turn from the discussion of packaging materials in general to a particular group of packaging materials: plastics.

People react to plastics in one of three ways:

- positive: plastics are the greatest thing invented since sliced bread;
- negative: plastics are the Devil's material that are destroying our planet; or
- don't think about it: they just take plastics for granted (and this is the vast majority).

But there is a fourth way we should look at plastics. Yes, plastics are brilliant in some circumstances, for example medical applications where hygiene and the ability to mould plastics into complex shapes is invaluable. But they have become ubiquitous, for creating both durable goods and packaging. Their use is out of control and they are seriously damaging our environment. We have to fundamentally rethink our approach to and use of plastics.

But it is also important to remember that as Paul Harvey says "Plastic itself is not inherently bad; it is the way in which we use and dispose of the plastic that makes it troublesome to the natural world".[1]

If we can find acceptable ways to manage unwanted plastic, we could continue to avail ourselves of its undoubted benefits. Plastic is not the Devil's material; it is the way we manage it when we no longer want it that is the problem.

DOI: 10.4324/9781003504757-10

The reason plastics have become ubiquitous is fourfold:

- they are easy to form into the product that is wanted, whether this is a durable product or packaging;
- there are many, many plastics to choose from meaning that there is always a plastic to do exactly what a manufacturer wants it to do;
- they are strong, light and hard wearing, good qualities for durable products and qualities that protect products where they are used as packaging; and
- they are cheap to produce.

From a manufacturer's point of view, plastics are great, which is why they have become so prevalent.

But the big, and I mean big, downside (and you probably do not realise this) is that once a product has been made out of plastic, it will be around **for ever**.[2] Plastics don't decay, break down or disappear. Yes, over very, very long periods of time (multiple decades or even hundreds of years), plastics degrade, but all they do is break down into smaller pieces of plastic (called microplastics). When this happens, they become even more dangerous as they can be ingested by animals and fish which we then eat and therefore we ingest the plastic.

Does it matter if we ingest plastic particles? Apparently it does. Medical research in 2022 found plastic particles in the blood of 22 volunteers. The most common polymers were: polyethylene terephthalate (PET); polyethylene (PE); polymers of styrene (e.g. rigid polystyrene, expanded polystyrene and others); and polymethyl methacrylate (PMMA) – perspex to you and me.[3] These are all high-volume, common polymers.

Okay, so the plastic particles were in the volunteers' blood stream, but was this a problem? And this is where the research grew really worrying. A second study discovered microplastics deep in human lung tissue. The researchers raised concerns that the sharp rigid particles of plastic may cause the same damage as asbestos fibres. Now that is a major health concern.

A little history

Before looking to the future, let's just understand where plastics came from and how our love affair with plastics began.

Alexander Parkes patented the first plastic "Parkesine" in 1855.[4] In 1869 John Wesley Hyatt patented Celluloid which was used to make, among other things, billiard balls, which were, appallingly, previously made from ivory.[5] In 1907 Leo H. Baekeland developed Bakelite, the world's first durable plastic.[6]

The mass production of plastic products started in the 1920s, but really started to take off in the 1940s, particularly during the war years, when plastics were increasingly used in military products.

But it was the post-war consumer boom in the late 1940s and 1950s that really caused us to fall in love with plastic. As Lucy Siegle says in her book *Turning the Tide on Plastic*, "Once we took to plastic, we fell under its sway with indecent haste. In just a few short years, a make-do-and-mend domestic culture passed down from one generation to the next had been turned on its head".[7] So not only did we welcome plastics into our lives with open arms, we destroyed a culture of manufacturing and repairing that used sustainable materials and methods. I find that thought so depressing.

The problem with plastics

By the late 1970s, environmentalists were starting to raise awareness about the toxicity of some plastics, such as PVC, which has been shown to be carcinogenic (which is why it is quietly being dropped as a packaging material, particularly for food). Some plastics are also known to leach toxins into food when they are in contact with fat or when they are heated, for example in a microwave oven.[8]

The two key problems with plastics are:

- the sheer volume of unwanted plastic "waste"; and
- the fact that so much of this unwanted material is not being properly managed and treated at the end of its useful life.

There is a much-quoted projection that there will be more plastics by weight than fish in our oceans by 2050;[9] what an appalling thought. So, we do need to talk about how we manage our unwanted plastics.

To quote Paul Harvey again "[Plastic] waste has become an ever-increasing problem, as we do not have a solution to the waste generated once the plastic is no longer of use to us. Plastic is discarded carelessly, haphazardly and irresponsibly all over the world and has created an environmental catastrophe that we are only now, as a civilization, waking up to".[10] I couldn't have said this better.

The solution to how to deal with unwanted plastics would appear to be recycling, but plastic recycling is problematic for at least four technical reasons and one economic one:

- there is a plethora of plastic types in use in consumer products and packaging, at least seven key polymers with many minor variants;
- separating plastics into the different polymers for reprocessing is not easy and the automation of this is limited;
- plus, multiple different polymers are often combined within one product or item of packaging, for example consumer bottles, such as for detergents,

are often made with caps of different polymers and the separation of these polymers can be expensive and technically challenging;

- plastics are very light so that transporting collected plastic products and then cleaning them is expensive; and
- oil (the raw material from which plastics are derived) is relatively cheap and so production of virgin plastic raw materials can be much cheaper than recycling.

Plastic currently forms a significant part of Household Waste (about 9% by weight in 2017, but much more by volume) and so we need to address its collection and treatment. This should form part of the mechanical treatment cycle and, as suggested above, this needs to be done by recycling.

The Plastics Pact

Led by WRAP, the plastics industry's Plastics Pact has the following four targets:

- Target 1: eliminate problematic and unnecessary single-use plastic;
- Target 2: 100% of plastics packaging to be reuseable, recyclable or compostable;
- Target 3: 70% of plastics packaging [is] effectively recycled or composted; and
- Target 4: 30% average recycled content across all plastic packaging.

Progress to date is as follows:[11]

- Target 1: 84% reduction in problematic and unnecessary plastic packaging since 2018, which appears to be good progress;
- Target 2: 70% of plastic packaging is recyclable – up from 66% in 2018, but this presumably means recyclable in theory only;
- Target 3: 50% of plastic packaging is recycled – up from 44% in 2018, but I challenge this number as it includes plastic "waste" sent overseas for so-called recycling; and
- Target 4: 22% average recycled content, up from 8.5% in 2018.

Some statistics

The following statistics are taken from the latest WRAP report on plastics packaging, published in 2021.[12]

In 2019, 2.3 million tonnes of plastic packaging entered the UK market. Of this 1.5 million tonnes was consumer packaging (63%) of which 68% was in the grocery sector, so 43% of all plastic packaging was for food and drink.

Plastic bottles accounted for 44% of the consumer plastic packaging, plastic film 21% and pots, tubs and trays also 21%.

The report cites a UK recycling rate for plastic packaging of 51%; however, more than half of this was achieved by exporting the plastic "waste", so the real UK recycling rate was actually only 22% (see my comments below on Household Waste exports).

In terms of Household Waste collection, in 2018/19 WCAs collected 572,000 tonnes of post-consumer plastic "waste", only 38% of the total consumer packaging entering the market.

Plastic film

Even though 99% of UK WCAs collect plastic for recycling,[13] only 17% of WCAs collect plastic film in their kerbside schemes.[14] About 16,000 tonnes of plastic film was collected by WCAs in 2019, representing a collection rate of only 7%.[15]

Plastic film accounted for nearly 16% of the plastics collected via kerbside collections in 2018.[16] Not all of this was packaging; bizarrely some of this was the disposable plastic sacks that some WCAs give to households to collect their recyclables (20% of the 16%),[17] so aren't they adding to the problem? Carrier bags accounted for a further 1.5% and the rest (11.2% or 56,485 tonnes) was made up of things like crisp packets, bread bags, chocolate bar wrappers and multipack bags. The different polymers used to make these packaging products were not identified (film is not marked in the same way as rigid plastic packaging), but plastic film is predominantly Low Density Polyethylene (LDPE) and polypropylene (PP).

I recycle as much as I can, practise reuse and try to minimise my consumption. And I have found that the proportion of my Household Waste that requires disposal has reduced to an extraordinary degree. I am fortunate in having a weekly food "waste" collection service provided by my WCA and by also separating out Dry Recyclables; what actually goes into my wheeled bin for disposal is both clean, very light and requires collection quite infrequently.

My Household Waste for disposal is collected on alternate weeks to Dry Recyclables, so it's collected fortnightly. But I often find there is so little in this wheeled bin that it is not worth putting out. This element of my Household Waste can easily be left quite safely for at least four weeks between collections, possibly even longer. This is one of the reasons why I advocate having wheeled bins for Household Waste for disposal that are smaller than at present. But what has truly surprised me is that the majority of the materials in this bin are plastic film. If we can find ways to make this plastic film recyclable, we will reduce substantially the amount of Household Waste being disposed of.

Some producers mark their film products with the OPRL "Recycle at larger stores". Some of the biggest grocery retailers have a small (and they are small and usually full) collection bank at the entrance to their biggest stores, for the collection of plastic film. These were originally introduced to collect unwanted plastic carrier bags, but most stores now accept other types of plastic film as well. I recently asked one major grocery retailer what happens to this collected plastic film. They told me they didn't know as the collection was outsourced, which goes to show how unimportant such collection was to them.

Such an approach is surely all about marketing or, dare I say, greenwashing? Okay, a very small percentage of the consumers shopping at one of the few stores that have such collection banks will collect and store their unwanted plastic film packaging (and it is in very large part packaging) at home. Then they have to remember to take it with them when they go shopping. But what about the majority of consumers who do not frequent these large retail stores or don't choose to take their unwanted plastic packaging film with them to the supermarket?

This cannot be seen as a viable method of collecting plastic film for recycling. As I say, if anything it is greenwashing. So what are we to do with all of our unwanted plastic film packaging? I'll answer this fully in Chapter 10 under "The materials that are actually recyclable", but for now I'll say there are two things we should do:

- producers should limit the number of polymers they use to perhaps just the three that are truly recyclable (rigid PET and HDPE, and flexible LDPE); and
- WCAs should include plastic film products in kerbside Dry Recyclables collection schemes.

But more on this later.

The markets for recyclable plastics

Recycled plastics are sold within a global market and because they compete directly with virgin plastics the prices achieved for recycled plastics are subject, among other things, to variations in the price of oil. The markets for recycled plastics are therefore volatile and to sell successfully, recycled plastics need to be as high a quality as possible.[18]

This makes absolute sense, but what doesn't make sense to me is why the UK exports "waste" plastic for reprocessing overseas (over half of plastics packaging sent for recycling is currently shipped abroad, about 500,000 tonnes (or 22% of plastic "waste" arisings),[19] yet this exported tonnage is included within the reported UK recycling figures).[20]

Why do we not reprocess all of our own collected post-consumer plastics? The answer is because there is insufficient reprocessing capacity in the UK. Again, why?

Well, this is down to economics; it's cheaper to export "waste" plastic for reprocessing to some countries such as China, Turkey and Malaysia, than it is to reprocess it in the UK. Again why? Because these countries do not operate to the high standards that exist in the UK, often employing poorly paid and unprotected people, sometimes children, to hand pick through plastic "waste" that is just dumped on open land (see the section "The future role of plastics" on p. 131).

The markets for plastic film are particularly problematic, with much of the collected film being exported, or if it is reprocessed in the UK it is into what is called Refuse Derived Fuel (RDF) which is burned as pellets as a fuel source (a form of recovery, not recycling).

One polymer that can be used to manufacture new **products** from unwanted plastics is Polyethylene (PE). This is because High Density Polyethylene (HDPE) and Low Density Polyethylene (LDPE) are thermoplastics and so, when heated, soften and can be extruded (squeezed through a die like toothpaste out of a tube) into useful shapes. When cooled, the material retains the extruded shape. So, HDPE and LDPE packaging, whether rigid or film, can be reprocessed into things like fence posts, street furniture or any other reasonable chunky item. But as Henry Ford famously said of the Model T Ford car, "You can have any colour you like, so long as it's black". This is because HDPE and LDPE packaging comes in many colours and the only way to make consistent products from this material is to colour it black.

The other attraction of manufacturing new products out of unwanted HDPE/LDPE is that PE can be used almost as a glue to hold pieces of other plastics together. Less recyclable plastics like polypropylene (PP) and hard polystyrene (PS) can be chopped into small pieces and suspended in a matrix of HDPE/LDPE thus bulking up the material and putting the less recyclable plastics to some use.

A new use that is being developed for unwanted PE is the fledgling plastic road application, where HDPE/LDPE is being trialled as a replacement for a proportion of the tar that is used to hold together the aggregate (stones) in tarmac. A perfect use for black plastic you might think. But as I've said several times already, you have to design-with-the-end-in-mind. What will happen to the plastic in these roads as they wear and when they eventually have to be replaced? Roads are re-surfaced by planing off the surface, i.e. grinding it into small pieces and then reusing the planings as aggregate in the new road surface. Are we not risking releasing tonnes of microplastics into the environment in ever reducing particle sizes, plastic particles that cannot be controlled or ever recovered?

Biodegradable plastics and why they are a bad idea

Given that so many people are now aware of the problem of plastic "waste" in our environment, producers are trying to find alternatives for some applications. For example, I've started to see carrier bags that are branded as compostable or "100% degradable". An example of the kind of labelling printed on such bags is shown in Figure 9.1.

Producers of biodegradable plastic films, such as bags or magazine envelopes, utilise one of two techniques to make the plastic degrade:

- additives are added to standard plastics to break down the long-chain molecules in the plastic polymers; or
- plastic-like materials are made from non-oil raw materials, such as potato starch or seaweed, as was the case of the recent Earthshot "Build a Waste Free World" prize winner.

But, the problem with the former is that by breaking down the long-chain molecules, all we do is turn the plastic film into smaller pieces of plastic, which as I've said can be even more damaging for the environment. These additives don't actually make the plastics break down into harmless bi-products. Plus the conditions required for the chemical breakdown to occur (it requires the presence of oxygen) are not those found in landfill sites or even in soil.

FIGURE 9.1 An example of the labelling of a compostable bag.

If, however, the plastic film is made from organic material, such as potato starch, then it might compost down into harmless bi-products if it is put into a commercial composting environment, but certainly not a home compost heap.

There are two serious problems with what is written on the bag in Figure 9.1.

First, many WCAs tell their residents to line their food "waste" caddies with newspaper, a purpose-made paper bag liner or a plastic bag. The reason that a plastic bag can be used is that when the food "waste" is sent for anaerobic digestion, all plastic bags are mechanically removed from the food "waste" before processing.

This means that if a compostable plastic bag is used to line the food "waste" caddy, it will be removed and disposed of prior to the processing of the food "waste". The equipment cannot differentiate between an ordinary plastic bag and a compostable one, so the bag will not be recovered. This completely negates the idea of having a compostable bag.

Indeed, in a survey of anaerobic digestion plant operators in 2019, WRAP stated that the majority of commercial site operators reported that "… compostable packaging was having an impact upon their business through increasing reject levels (and disposal costs) and causing problems in the anaerobic digestion process because it did not break down".[21]

Second, if compostable plastic bags are mixed with standard plastic bags, the compostable bags are actually a contaminant in the plastic recycling process and can ruin a batch of recycled plastic. This is why the producer of the bag says do not mix this bag with plastic recyclables.

I suggest that if producers are going to use "compostable" plastic bags or film, these should only be put into WCA garden "waste" collection bins and not onto home compost heaps. However, they are very likely to be rejected at the start of the composting process (see Annex I), plus getting this message across to the householder will be problematic and many householders will not do what is required. This will mean that at best, the "compostable" plastic bag will not be recovered or, at worst, it will contaminate plastic film sent for recycling, causing the rejection of batches of recycled plastic film.

Far better to not use "compostable" plastic film at all. I know this sounds defeatist, but we've got to keep the messaging simple and this is just too complicated and the risks of failure too high.

And I'm not alone in saying this. In his excellent book *How Bad are Bananas?*[22] Mike Berners-Lee says "Biodegradable plastic packaging is worth a mention because it can be a well-intentioned disaster area. It sounds great, but if you send it to landfill it rots down and emits methane and if you throw it into the recycling bin it can ruin the entire batch". So, let's forget compostable plastic.

And let's not overlook the fact that compostable plastics are a form of recovery, not recycling. When the compostable plastic is no longer wanted, it is in theory recoverable within the organic recovery cycle, i.e. composting. So, this is not a circular solution.

Just as an aside, I recently came across something called "Stone Paper". I looked this up on Wikipedia and found the following: "Stone paper products, also referred to as bio-plastic paper, mineral paper or rich mineral paper, are strong and durable paper-like materials manufactured from calcium carbonate bonded with high-density polyethylene (HDPE) resin … The production of stone paper uses no acid, bleach or optical brighteners. It can be recycled into new stone paper, but only if recycled separately at dedicated civic amenity sites". Hah!

Whilst there appear to be environmental benefits to this paper, as the article goes on to say "It is not biodegradable or compostable, but is photo-degradable under suitable conditions. It consists of roughly 80% calcium carbonate, 18% HDPE and 2% proprietary coating".[23]

Not only is this product unsuitable for kerbside recycling or organic recovery, but if mixed with other paper for recycling, it would contaminate the paper being recycled. This is thus another example of just-because-you-can-doesn't-mean-you-should. Where's the design-with-the-end-in-mind? As with compostable plastic, I think this is a product we should do without.

Plastics labelling

You will no doubt have noticed that many plastic items, particularly hard plastic packaging, are marked with a symbol comprising a number and letters to identify the type of plastic from which the packaging is made. These are shown in Table 9.1.

It has to be said that marking plastics with the symbols shown in the first column of Table 9.1 is pretty pointless. The vast majority of us don't know what they mean and so don't look at them and anyway they do not guarantee that the plastic is recyclable or is recycled. Given that discarded plastics, if sorted, are sorted in a MRF using automated techniques, who is actually reading these symbols? The only benefit they bring is if you are concerned about plastics, you might read the symbol and decide **not** to buy the product if it is made from or packaged in a plastic that you know is not recycled, such as PVC.

As well as explaining the symbols used to identify the different plastic polymers in use, Table 9.1 also shows the relative mix of the different polymers present in Household Waste. Two polymers dominate, PET and HDPE (representing about 62% of all plastics in Household Waste), both of which can be successfully reprocessed at an acceptable financial cost.

TABLE 9.1 Plastic material labelling symbols and their meaning

Recycling symbol	Plastic name	Plastic polymer	Percentage of kerbside collected tonnage* excluding film	Whether currently, widely recycled	Typical uses
PET	PET or PETE	Polyethylene Terephthalate	40.3%	Yes	Water and soft drinks bottles, biscuit trays
HDPE	HDPE	High Density Polyethylene	21.6%	Yes	Milk and juice bottles, ice-cream tubs, shampoo and detergent bottles
PVC	PVC	Polyvinyl Chloride	0.1%	No	Cosmetics' containers, commercial cling film
LDPE	LDPE	Low Density Polyethylene	N/A	No	Squeezable bottles, cling film, bin liners, carrier bags
PP	PP	Polypropylene	10.2%	No	Microwaveable trays, ice-cream tubs, crisp packets
PS	PS	Polystyrene	0.5%	No	Yoghurt pots and CD/DVD cases, plastic cups/cutlery
EPS	EPS	Expanded polystyrene	1.9%	No	Take-away cups and clam shells, meat trays, protective packaging, e.g. TVs
Other	Other plastic	Unidentified	0.3%	No	

TABLE 9.1 (Continued)

Recycling symbol	Plastic name	Plastic polymer	Percentage of kerbside collected tonnage* excluding film	Whether currently, widely recycled	Typical uses
81 C/PAP	Combined paper and plastic or aluminium	e.g. Tetra Pak type cartons	N/A	No	Milk, fruit juice, soup cartons

Note: * WRAP, "Composition of plastic waste collected via kerbside", October 2018, derived from page 12.

The future role of plastics

There have been two recent events that will change the way plastics are viewed and recycled in the UK.

First, in his budget statement on 11 March 2020, the then Chancellor of the Exchequer, Rishi Sunak, announced that he was introducing a tax on plastic packaging that does not incorporate at least 30% recycled content.

Second, on 21 April 2020 the price of crude oil went negative for the first time ever. Oil producers couldn't give the stuff away and had to pay people to take it. It is an understatement to say this was unprecedented. Obviously, the oil price recovered, but it's a sign that things could change.

But then third, as cars and other vehicles increasingly become electric, the demand for oil for petrol and diesel is going to fall. As renewable forms of energy generation increase, such as solar panels on houses and wind farms, the demand for heating oil will also reduce. What is clear is that the demand for oil **as a fuel** is going to go down. But oil extraction and production is a major industry in many countries and exporting oil is a major part of these countries' national economies, just look at the Middle East and the USA. So, what are these countries going to do with the oil they are inevitably going to continue to produce, if the demand for oil as a fuel reduces?

Well, one of the things they can do is push more oil into the manufacture of plastics. According to Paul Harvey "In a report by the International Energy Agency in 2018, it was modeled that plastics will be the largest growth industry for oil and gas refining products up to the year 2050".[24] Many people think we as consumers will face a tsunami of new plastic as producers are able to access more and cheaper plastic for their products and packaging. As consumers we can do our best to resist by simply refusing to buy such products and plastic packaged products and lobbying producers, retailers and, in particular, Government, to deter any such move.

The future of consumer plastics

In the introduction to this section, I said there is a fourth way we should look at plastics. This is to only use plastics where they add real value and only provided, and this is the big caveat, that they can be successfully and easily recycled back into virgin equivalent plastic that can be used to make new plastic products. We would then have circular, closed loop recycling and the plastic would go round and round, virtually for ever.

Currently, the recycling of plastics is a mechanical process: collected, unwanted plastics are sorted, washed, shredded and turned into pellets, which can be used to replace virgin plastic. But this is a relatively crude process and has limitations in terms of sorting different polymers to produce clean streams of single polymers. And the costs of such recycled plastics are relatively high when compared to virgin material.

Encouragingly, a lot of research is being undertaken to try to develop ways to chemically or biologically recycle unwanted plastics. This approach breaks down the unwanted plastics into basic polymer building blocks that can be built up again to produce whatever plastics we want. Some approaches use chemical processes, others use enzymes or bacteria to break down the unwanted plastics. One recently reported project is that being developed by Epoch BioDesign,[25] run by Jacob Nathan, which is using leading-edge scientific techniques to search for a solution. The article cited below said "Within a few years he [Nathan] hopes to have transformed this into an industrial process, producing chemicals at scale". I hope he does, but such reports of scientific progress have been around for quite a few years and we seem to be no nearer a solution. But as well as solving the biochemical challenges, there is always the question of cost. If recycling costs more than virgin extraction, then virgin extraction will always win. This is why there may need to be intervention by Government to create a level playing field between recycled and virgin raw materials, or, better yet, to tip the cost balance in favour of recycled materials.

But whilst such research projects give us hope of a solution to the problems of unwanted plastics, they are today only that, a hope. So, to make progress today, we need to accept that the reprocessing of plastics will continue to be mechanical and we will have to address the collection and sorting processes within this context.

This would limit the use of plastics to those plastics that can be easily and effectively mechanically recycled and I suggest these are PET and PE.

Conclusions on our use of plastics

I started by quoting Dr. Paul Harvey who said, "Plastic itself is not inherently bad; it is the way in which we use and dispose of the plastic that makes it troublesome to the natural world". So, we have to fundamentally rethink our approach to the use of plastics and their post-use treatment.

Whilst reuse of no longer wanted plastic packaging should be our first choice, in reality this will have a limited impact, so it is more realistic to focus on recycling plastics. There is a significant level of research into the potential chemical and biological treatment of unwanted plastics, but commercially viable solutions are years away and their cost is unknown, so we have to focus today on mechanical recycling.

Given the difficulties of sorting and separating different plastic polymers, particularly plastic film, in an automated MRF, the only current solution is to limit the number of polymers used by producers in their products and packaging. I suggest that with regard to packaging, this should be limited to PET and HDPE bottles and LDPE film packaging, and if this were done then plastic film could and should be included in kerbside recycling collections.

Notes

1 Dr Paul Harvey, *The Plasticology Project*, Indie Experts, 2022, p. 170.
2 The only plastic that no longer exists is that which has been incinerated or recycled and this is a tiny proportion of the plastic that has been manufactured since the 1960s.
3 Dr Paul Harvey, *The Plasticology Project*, p. 104.
4 Lucy Siegle, *Turning the tide on Plastic*, Trapeze, 2018, p. 36.
5 Lucy Siegle, *Turning the tide on Plastic*, p. 40.
6 Imperial Chemical Industries Ltd, *Landmarks of the Plastics Industry 1862–1962*, The Kynoch Press, 1962, pp. 13–25.
7 Lucy Siegle, *Turning the Tide on Plastic*, Trapeze, 2018, p. 43.
8 Heather Rogers, "A Brief History of Plastic", The Brooklyn Rail, May 2005, https://brooklynrail.org/2005/05/express/a-brief-history-of-plastic, accessed 20 March 2023.
9 Ellen MacArthur Foundation, "The New Plastics Economy: Rethinking the Future of Plastics", World Economic Forum, January 2016.
10 Dr Paul Harvey, *The Plasticology Project*, Indie Experts, 2022, p. 10.
11 WRAP, "The UK Plastics Pact Annual Report 2021–2022", December 2022, p. 4.
12 WRAP, "Plastics Market Situation Report", 2021, pp. 6–11.
13 WRAP, "Composition of plastic waste collected via kerbside", October 2018, p. 6.
14 WRAP, "Plastics Market Situation Report", 2021, p. 12.
15 WRAP, "Plastics Market Situation Report", p. 13.
16 WRAP, "Composition of plastic waste collected via kerbside", October 2018, p. 13.
17 WRAP, "Composition of plastic waste", p. 24.
18 WRAP, "Composition of plastic waste", p. 8.
19 WRAP, "Plastics Market Situation Report", 2021, p. 11.
20 Ian Tiseo, "Plastic waste trade in the United Kingdom – statistics & facts", statista.com website, accessed 22 November 2021.

21 WRAP, "Anaerobic Digestion and Composting Industry Market Survey Report", April 2020, p. 5.
22 Mike Berners-Lee, *How Bad are Bananas?*, Profile Books, 2010, p. 81.
23 "Stone paper", https://en.wikipedia.org/wiki/Stone_paper, accessed 28 July 2023.
24 Dr Paul Harvey, *The Plasticology Project*, Indie Experts, 2022, p. 175.
25 Ben Spencer, the *Sunday Times*, "Young scientist's quest to fashion old crisp packets into new shirts", 12 February 2023.

10

WHAT WE NEED TO DO DIFFERENTLY

Right back in Chapter 1, I set out the problems we face from consuming the Earth's limited natural resources at an unsustainable rate and then trashing the planet by discarding materials when we no longer want them.

We have to change and we are going to have to do a few things, well, quite a lot of things, differently. So, in the next two chapters I want to talk about:

- what needs to be done differently and by whom (this chapter); and
- what we could realistically achieve (the benefits) if we changed (Chapter 11).

I've already said that the broad solution to our problems is to move away from our current approach of linear, single-use consumption to one of managing our resources in a much more circular way. Whilst this is relatively straightforward for metals, glass and plastics, there are some materials for which this isn't possible. But some of these materials, such as wood, paper, cardboard and natural textiles, could be sustainably replenished provided there is sufficient land available to grow the required trees and crops.

For those materials that cannot be treated in a circular or sustainable way, we need to consider simply designing them out of our products and packaging. Sometimes this means simply going back to a previous packaging format that was recyclable. For example, take chocolate bars. In the past, many chocolate bars were wrapped in recyclable aluminium foil, with a recyclable paper wrapper for printing on. Whilst a very few still are, the majority of chocolate bars are now wrapped in non-recyclable plastic film

DOI: 10.4324/9781003504757-11

packaging. The change was presumably for the benefit of the producers, but has made the packaging of chocolate bars non-recyclable. Why don't we go back to what was a sustainable packaging solution? So instead of trying to come up with ways to recycle inappropriate packaging, let's redesign the packaging to make it recyclable.

Whilst this might all sound like a big ask, I truly believe we can achieve it by all of the players involved (not just you and me as consumers) making many relatively small changes to how we currently consume the Earth's resources and to how we manage Household Waste. But it will require all the players in the Circular Supply Chain to do their part and to act with a common goal and it will require the creation and operation of new, local and regional recycling and recovery facilities.

As I said right at the beginning of this book, the starting point for this change is to stop seeing what is discarded by householders as "waste" and to start seeing it as useful materials, ready for a second, third … infinite life, and to put in place the infrastructure and attitudinal changes that will transform how we manage the Earth's limited resources.

I've also said that we need to manage "the problem" by starting at the beginning of material use, by designing products and packaging so they are suitable for reuse, repair, recycling and recovery, instead of expecting local authorities to deal with the mixture of treatable and untreatable unwanted materials, after they have been discarded. I've called this design-with-the-end-in-mind and this is the starting point for what has to change.

The Materials Hierarchy

As I proposed in Chapter 3, instead of the Waste Hierarchy I think we should talk about the Materials Hierarchy, comprising the following methods of treatment in decreasing order of attractiveness, what I call the Three Rs:

- **reduce** Household Waste by:
  - reducing our consumption;
  - increasing reuse; and
  - increasing product repair;
- **recycle** where possible what can't be reused or repaired; and
- **recover** what can't be recycled as:
  - organic material; or
  - energy.

If all else fails then we will dispose of the small remainder to landfill as a last resort, but it has to be the very last resort.

Cost and the Materials Hierarchy

Having talked earlier about the dominant economic model, that of capitalism, I'd now like to talk about a more practical aspect of economics, that of financial costs.

As Tom Szaky says in his book *Outsmart Waste*,[1] "In the end all waste can be reused, upcycled or recycled ... The challenge in all of this ... is one of economics". By "economics" he really means cost. So, I want to take a moment here to reflect on one of the major constraints to change and to consider how best we can address it.

In the end it is the consumer/householder who pays for the treatment of our unwanted materials, both through the item's initial purchase price (if it is to be made in a more sustainable way it is likely to cost more) and through our Community Charge to cover the costs of our Household Waste collection and treatment. This is absolutely as it should be. If we want to consume a product, we should be prepared to pay for its collection and treatment once we have finished with it.

Now, producers and retailers worry about the initial purchase price and try to keep this as low as possible whilst still making a profit, and local authorities are concerned to minimise Community Charge rates, in order for their members to get re-elected. Everyone wants to keep costs down (including of course consumers), so most decisions are to go with the cheapest cost option.

As I've already said, in terms of products this has increasingly meant a move away from reuse and repair and to the selection of materials that suit the producer and not necessarily recycling.

But much, much worse, it has led to the idea of it being okay to make products and packaging disposable, because this is cheaper than treating them once they are no longer wanted. This was particularly true when landfill was so cheap. Back in the 1950s and 1960s when the idea of products being disposable for increased consumer convenience really started to take hold, landfill was very cheap. In the 1970s this idea of "just bin it when we no longer want it" was well established and so it wasn't until the 1980s, when tighter environmental regulations started to increase the costs of landfill and then the Landfill Tax was introduced that landfill started to become a more expensive option.

The relative costs of the Materials Hierarchy

It was only when the cost of landfill started to rise significantly after the turn of the last century, that WCAs and WDAs really started to become interested in recycling as a more cost-effective alternative to landfill. As the Landfill Tax increased year-on-year, more and more WCAs and WDAs turned to kerbside recycling and recovery of Household Waste as a cheaper form of treatment than landfill. So, the Landfill Tax did its job and we as householders all paid more for the collection and treatment of our Household Waste and, to be fair, we didn't really notice. So, a good job done? Well yes and no.

The tighter environmental regulation of landfilling and the introduction of the Landfill Tax has certainly led to widespread kerbside collection for recycling and organic recovery becoming the norm for a limited range of materials, so that's the good part. But, because, currently, the infrastructure for recycling and recovery is limited and because producers and retailers haven't bought into making changes to facilitate the large-scale recycling of Household Waste, WDAs have turned to EfW incineration as the cheaper alternative to landfill, rather than pushing recycling and organic recovery to its full potential. Again, financial cost has come into play and we see the Law of Unintended Consequences[2] taking hold.

As the Landfill Tax has risen, it has not only made recycling and organic recovery more financially viable as an alternative to landfill, but it has done the same for EfW incineration. The waste management industry spotted this opportunity to replace landfill with EfW incineration and when faced with the continuing need to dispose of large quantities of Household Waste that couldn't be recycled, WDAs chose EfW incineration as the best financial solution, compared with the increasingly expensive option of landfill.

But an EfW incinerator is a big financial investment so, understandably, the operators of such plants require long-term guarantees of significant quantities of Household Waste inputs to make their investment pay off. So now we are back to the situation that existed in the 1980s, where there is a form of disposal (and, as I have explained in Chapter 5, whilst EfW incineration is a method of recovery, it is right at the bottom of the Materials Hierarchy) that makes it very difficult for WDAs to increase the recycling of Household Waste. I explain this more fully later in this chapter, but for now let's just park this as a major issue of concern.

If I'm brutally honest, it has been cost that has driven the changes in how we collect and treat our Household Waste, rather than concern for the environment. Yes, the environmental benefits of recycling over landfill have been recognised and trumpeted as the right thing to do, but change has only really come about in order to minimise treatment cost increases.

I said earlier that the cost issue we face is one of **relative** cost. Put simply, what we have to do is make changes to ensure the costs of different collection and treatment options reflect the priorities of the Materials Hierarchy. Given that all players in the Circular Supply Chain will generally select the lowest cost option, we need there to be an inverse relationship between environmental benefit and financial cost, that is, the greater the environmental benefit, the lower the cost. So, for Household Waste this would mean that:

- reduction which has the highest environmental benefit should be the lowest cost option (which it is);
- recycling will cost more, but will deliver high levels of environmental benefits;
- organic recovery which will deliver lower, but still significant, environmental benefits should be a similar cost to recycling; and
- EfW incineration, which delivers the lowest environmental benefit should cost more than recycling and recovery (potentially through an Incineration Tax on the treatment of Household Waste – see later).

And, of course, landfilling, which delivers no environmental benefit at all, should cost the most, and through the Landfill Tax we already have the mechanism to achieve this.

By aligning the economic costs of different treatment options with their environmental benefits, we will start to drive the right behaviours, by all players.

As I've said, it's actually economics that drives change, not environmental concerns. Therefore, we have to change the economics so that they drive the environmental behaviours we so desperately need and this is where the Government can play its part. Just look at the positive effects that the Landfill Tax had on the recycling of Household Waste. But, as I will argue in a moment, we now need our national Governments to introduce an Incineration Tax, certainly for Household Waste, to drive the methods of treatment up the Materials Hierarchy and away from EfW incineration.

Looking through the wrong end of the telescope

And if we stop and think about it, we've been approaching the development of recycling and organic recovery from entirely the wrong direction. It has been WCAs and WDAs who have been driving the recycling and recovery of Household Waste, because the collection and treatment of Household Waste is their responsibility.

But it is also true that they have always come at it from an understandable perspective of minimising their costs, and not so much from a concern for the environment. And because they have been looking only at the collection and treatment of unwanted products and packaging, they have been addressing the symptoms of the problem, not the underlying causes.

The underlying constraints to our ability to recycle and recover products and packaging is that they are, in the main, not **designed** for recycling and recovery. Products and packaging are designed for consumption and for lowest cost manufacturing; what happens to them after the consumer no longer wants them is not the producer's or retailer's concern or cost. But if we are to make the progress I believe is possible, we need producers and retailers to take responsibility for designing products and packaging so they are suitable for recycling and recovery. I'm not saying that producers and retailers have to bear the cost of recycling and recovery (they wouldn't anyway, they'd simply pass it on to the consumer), but they do need to take responsibility for *facilitating* recycling and recovery through appropriate design. I'll say more about this in a moment.

That said, there are many people who have argued in the past that producers should pay the cost of treating their products at the end of their useful lives (whether it is a durable product or packaging). The argument is that producers make these products and packaging and so should take responsibility for them at the end of their useful lives. This is why this concept is called "Producer Responsibility".

This sounds like a fair argument, but it misses two key points:

- first, a producer only manufactures products that consumers want (they simply don't make what doesn't sell), so isn't it actually the consumer who should be responsible for the product or packaging at the end of its life, because they are the ones who bought and "consumed" it; and
- second, if producers were to have to pay for the treatment of their products at the end of their useful lives, they would simply add this cost to the purchase price of the product, so again it would be the consumer who would actually pay.

And on this second point, should it not be the person who benefits from the use of the product who pays for it and its packaging's ultimate treatment; isn't this all part of being a consumer? It's just that at the moment we as consumers do not pay for the treatment or disposal of our unwanted materials on a product-by-product basis, but we do pay via our Community Charge which covers the costs of recycling, recovery or disposal.

To come back to where I started this section on cost, as Tom Szaky says "In the end all waste can be reused, upcycled or recycled … The challenge in all of this …is one of economics". But as I have tried to explain, it is not just about economics (or rather costs), it's about **relative** costs, that is, which is the cheapest treatment option. As I said in Chapter 3, what we have done to date in terms of recycling and organic recovery has been to address the easy stuff. To go further we must make some fundamental changes and these will cost money, which means the consumer/householder will have to pay more through the item's initial purchase price and through their Community Charge.

If we are to increase the levels of recycling and organic recovery to what I think are possible, the methods of collection and treatment will cost more financially, but will deliver the environmental benefits that are vital to the health of our planet. So, in fact we face two challenges:

- how to increase the levels of recycling and organic recovery of Household Waste; and
- how to do this, whilst maintaining or not significantly increasing the costs to the consumer/householder.

Fundamental or incremental change?

Do these changes need to be fundamental or incremental? I like to think we can achieve what is needed through relatively small, but still significant changes.

Designing products and packaging to be recyclable or recoverable is not rocket science. It's more about having the right attitudes and fresh thinking. Yes, there may be some cost increases, but equally, there may be cost savings. And it is certainly true that the first mover producers will reap a significant benefit through increased sales if they get their marketing right, but not by greenwashing.

Changing our collection systems also does not need to be a fundamental change, indeed I advocate making it simpler for the householder. And the required treatment methods already exist and are in use, we just need more of them. So, I'm talking evolution, not revolution.

Having talked about cost as the key driver for change, let's now look at what needs to be done differently and by whom, taking each of the Three Rs in turn.

Reduce Household Waste

We should reduce Household Waste in four ways, by:

- reducing our consumption;
- increasing packaging reuse;
- increasing indirect product reuse; and
- increasing product repair.

Taking each one in turn …

Reduce Household Waste through reduced consumption

We all need to buy less stuff

A large part of the solution to our Household Waste problem is not about how we manage "waste", but about how we avoid creating it in the first place. And the way to achieve this is through changing how we, you and I, consume. Yes, this is a huge and challenging task, given that we live in a consumer society, but using a product or packaging once is simply not enough.

I've already said that I'm not naïve enough to think that we can radically reduce our consumption; buying stuff is too deeply engrained in our culture and is fundamental to our global economic model. Producers and retailers earn their living from us, the consumers, buying their products.

But it's blindingly obvious that, because we live in a consumer society and are bombarded by advertising, we're all buying too much stuff:

- some stuff we need;
- some stuff we really want, but don't necessarily need; and
- some stuff we just buy for the sake of buying something, anything, because it feels right in that moment (but the feeling usually doesn't last).

It's the last of these I would like to challenge.

William Morris, the leader of the Arts and Crafts movement famously said, "… have nothing in your house that you do not know to be useful, or believe to be beautiful". This chimes with the first two above reasons for buying something, but challenges the third. But I also accept that not everyone's definition of "useful" and "beautiful" will be the same.

We all need to buy more carefully

In terms of our own consumption, without wanting to sound preachy, we need to try to buy fewer products, whether these are clothes, household goods, food we just don't get around to eating or just stuff, and we need to choose products that are repairable and which are wrapped in packaging that is reuseable, recyclable or just not there (not everything needs to be packaged).

Here I'm talking about an attitudinal change to how we consume, not to stop consuming, but to think more carefully about how and what we consume. Three examples of this are impulse buying, desperation buying and emotional buying.

Impulse buying

We've all bought things on impulse and then regretted it. To paraphrase an old saying: "buy in haste and regret at leisure". Again, without wanting to sound as if I'm preaching, when tempted by a product that you're not totally sure you should buy, ask yourself "do I really need this or am I just buying it for the sake of buying it?" If we're honest, we all suffer from retail-therapy temptation from time to time. A good way to address this buying temptation is to take time out; walk away and go back later to buy what you thought of as your "must have" item, having thought about whether you really need it. And think "need" rather than "want"; do I really **need** this or do I just think I would like it. I'm not asking you to turn into a puritan or be a killjoy, just think before you consume.

Desperation buying

We've all been here; you need to get a birthday or Christmas present for someone and can't find anything really suitable and you're running out of time. So, you end up buying something, sometimes anything, to give them, without thinking enough about "will **they** really want this; do **they** really need it?" Desperation buying can leave the recipient with stuff they don't really want or need, but they now have the problem of what to do with it.

Again, I don't have the answer to this, except to say, please think hard about what you are going to give someone, from their perspective and not just your own desperation to buy something for them. I'd also suggest giving yourself enough time to shop; don't leave buying that gift until the last minute, when desperation buying is most likely to kick in.

Emotional spending

I've alluded to this already under the above heading of "Impulse buying", but what I'm talking about here is a concept particularly linked to online shopping.

According to Sirin Kale in a Guardian article in 2021,[3] emotional spending hit a high during the Coronavirus pandemic. She says: "The pandemic prompted a frenzy of online spending ... Most of this shopping was due to boredom". We're back to the dopamine hit that soon fades, as I talked about in Chapter 7. Kale says "Buying something online 'creates a small moment of joy, but it never lasts long' ... 'You realize there's nothing you really needed' ... 'you just got sucked into the moment and the high'". She goes on to say "The reason we are buying so much online is simple: we are online more than ever. It's like sitting in a pub all day and trying not to drink ... People are online all day and social media is full of things to buy. And the providers of social media are very skilled in persuading us to buy".

Buying more carefully

As consumers, we need to buy more thoughtfully. To quote the title of a book by Glenn Adamson[4] we need to buy *Fewer, Better Things*. This is a great maxim to live by. But I'd like to appropriate his title to mean we need to try to consume less and only buy good quality products that are made from sustainable materials and that are repairable and ultimately recyclable, and which are wrapped in packaging that is reuseable or recyclable.

But as Glenn Adamson acknowledges, not everyone necessarily has the means (or the motivation) to do this. He says "This model of sustainability through high-quality things has the unfortunate reputation of being elitist. Even more unfortunately, that reputation is entirely accurate. Few people can afford to fill their homes with finely crafted objects; insisting on excellence as our best pathway to ecological balance is simply impractical".[5]

Whilst I agree in large part with his conclusion, that we can't expect everyone to buy fewer, better things, we should still view this as an objective. Certainly, we should find ways to encourage everyone to buy fewer things. We need to find ways to discourage shopping for its own sake and buying stuff just for the sake of instant, but short-lived gratification (I refer you to my comments on impulse buying and emotional buying above and on Fast Fashion below).

My message however is very simple: we can avoid generating Household Waste by buying less stuff and so have less to throw away. What we have to focus on is reducing consumption rather than reducing "waste".

Buy, borrow or rent

But let's take a step back. Consumption is all about ownership. I buy something and it's then mine. This feeling of ownership is very important to us for all sorts of reasons, some good, some less so. But in a world of limited resources, do we really need to own all of the things we currently buy? Could we perhaps borrow or rent items instead? For example, do I need to own a garden shredder (which I don't, by the way), something that I would use once or twice a year and the rest of the time it would sit in my garden shed, gradually deteriorating until it got so bad I would have to chuck it out? Couldn't I, shouldn't I, simply hire one or borrow one from a neighbour, when I needed one?

Too often, we buy when we could borrow or rent. I'm not suggesting you go as far as Greta Thunberg who has said she isn't going to buy any new clothes: "Clothes? The worst-case scenario I guess I'll buy second-hand, but I don't need new clothes. I know people who have clothes, so I would ask them if I could borrow them or if they have something they don't need any more … I don't need to buy clothes I don't need, so I don't see it as a sacrifice".[6]

If you think about it, you've probably got more clothes than you need, so why buy more? Well, the answer is obvious; we all like to look good and a new item of clothing can give us a lot of pleasure. So, I'm not saying don't buy any new clothes, I'm simply saying please think before you buy and buy carefully and if you can, buy less.

We already accept that we will rent some items, because it would be silly not to. What city dweller wants to buy an expensive bike or an e-scooter that they will use only occasionally, when they can hire one and pick it up and return it conveniently, all arranged via a mobile phone. Many apparently. So, if this applies to products such as e-scooters, could it apply to other products?

Some producers are starting to think about radical, alternative models of consumption. For example, MUD Jeans is renting jeans to their customers instead of selling them.[7] When the denim is worn out customers return the jeans to MUD which then repairs or recycles them. Sounds great, but will it work? Will consumers want to rent rather than own their jeans, and will they and how will they return them when they no longer want them? These are early days, but it is encouraging to hear of a producer who is giving consumers an alternative way to deal with their products when they no longer want them. But this is also a really good example of us needing both the infrastructure to facilitate new behaviours and the attitudinal shift that will be required to make it happen. It's a big shift, and can it be made to happen at scale? Personally, I have serious doubts.

So that's about clothes; at the other end of the spectrum is cars. Do so many of us actually need to own a car? Apparently, the typical car spends 90% of its life sitting doing nothing except deteriorating,[8] or as much as 96% according to the RAC.[9] Is there an alternative? Obviously the answer to that is yes, as many of us don't own cars and instead use public transport. But there is also the rental sector. Many people only need a car occasionally and so when they need one, they rent one. Or there is another form of car rental – taxis. But car ownership is a funny thing. A lot of it is about status and about choice, having exactly what you want and it's there whenever you want to use it.

But recently there has been another change which has affected our attitude to ownership and that is the Covid-19 pandemic. Understandably, people do not want to share things with other people who might have the virus. So the use of public transport, such as trains or commercial transport, such as planes, fell off a cliff. At least if you own your own car, you know who has and who has not been in it, so you feel safer. Whilst I could advocate sharing more things like garden shredders and other DIY equipment, sometimes something comes along that pushes us in the exact opposite direction, like with cars.

But if you do decide to buy, then please choose products carefully and think about what you will potentially throw away: the packaging certainly and ultimately the product, unless it is consumable, reuseable, repairable or very long lasting. Which leads me to …

Buy-with-the-end-in-mind

We should always think about what will happen to the stuff we buy, after we have finished with it. This is what I call "buy-with-the-end-in-mind". When we are thinking about buying a product, we need to consider what will happen to that product and its packaging, once we have finished with it. This applies as much to food as it does to consumer products. Obviously with food we consume the product, but we have to think about what will happen to the food's packaging and buy the brand that uses truly reuseable or recyclable packaging. This is the first big change for us as consumers.

We cannot go on just buying whatever catches our eye; we have to think as we buy and buy more wisely. And this applies in spades when we come to what is called "Fast Fashion".

Fast Fashion

Fast Fashion is the term that has been coined to describe cheap, short-life clothing that has gripped a generation of clothes buyers and has become what is to many people a societal addiction.

In her book *How to Break Up with Fast Fashion*.[10] Lauren Bravo describes Fast Fashion as being characterised by two things: low prices and relentless pace. And I would add two further characteristics: relentless marketing and peer pressure. As Lauren says "... newness is everything. The pursuit of newness is a human instinct. But in recent years we've supercharged our pursuit of novelty and with it created a monster. A great, mutant monster with one hand in our wallet at all times. Fast Fashion".[11]

Lauren says "The pursuit of newness is a human instinct"; but is it? Has it always been or is it just part of our more recent consumer addiction? You may challenge this last description of our consumerism, but Lauren goes on to admit that "Throughout my life, clothes have been my passion and the acquisition of clothes my most devoted hobby ... I have danced out of shops cradling a carrier bag like a newborn baby ... Of course the One True Jacket ends up relegated to a lower peg a few months later, because you're bored and a button fell off and you had your head turned by another sexy piece of outerwear".[12] Can you honestly say you have never felt this way about some product that at the time of buying it, you felt you had to have it, only for the shine to wear off after a few months?

Fast Fashion is the opposite of how previous generations viewed clothes (and indeed most consumer products). Fast Fashion producers know that their consumers will fall out of love with their most recent purchases in an indecently short period of time; they know this because they created and sustain this situation. Because fashion-driven consumers keep buying so much clothing; over 1.1 million tonnes of clothing was bought in the UK in

2016[13] and 300,000 tonnes of "used" clothing was disposed of (that's 23% of all textiles and 1% of all Household Waste).

Another example of rampant consumerism, driven by relentless marketing and peer pressure is mobile phones. This, coupled with rapid technological developments drive consumers to want the latest, best, most advanced mobile phone on offer, only to find six months later that there is a better, more advanced, "must have" model now available. And I have to say that not only the purchasing of mobile phones has become an addiction, so has their use and not particularly for making calls, rather for accessing social media, music, apps and data. That's a whole other subject, but is relevant here as we have to change the way we view such devices and move away from Fast Phones as well as Fast Fashion.

Basically, we need to buy less stuff, which I know isn't always possible or necessarily desirable. As Lauren Bravo says "I'm a terrible person just like you're a terrible person … and nearly all of us as privileged folk who use and abuse the world's resources on a daily basis to feed our daft desires, are terrible people in unique and multifarious ways. It is hard to exist as a human in the developed world without sometimes being terrible. Or at least that's how it feels".[14] However, as Lauren goes on to say "[but] … you don't get far with guilt I find … 'Just buy less!' isn't a very effective message when everything else is still screaming 'BUY MORE!'"

This is why we need to change our approach, not just to how we manage Household Waste, but fundamentally to consumerism. I've already said that, ideally, we should all consume less, but I'm not naïve enough to expect this to happen in a big way or overnight. However, we can and do need to consume differently, starting with buying-with-the-end-in-mind. And we need producers and retailers to change their sales and marketing approaches. They have to become more responsible and stop pushing the damaging message "Just buy more".

And when buying, we consumers need to buy products with minimal packaging or packaging that can be reused or recycled, and to then reuse and recycle as much as we can. So please buy products packaged in cardboard, glass or metal and don't buy plastic (particularly plastic film) or composite packaging.

Ideally the first change I'd like to see to achieve Household Waste reduction (the highest priority in the Materials Hierarchy) would be a major shift in how much we consume. Is this likely to happen? Sadly, I think not. Too many of us are shopaholics and producers and retailers have a vested interest in this continuing. If we accept that levels of consumption are unlikely to drop substantially, we have to find ways of making consumption more sustainable.

Let's turn now to a different form of "waste" reduction, that of reuse.

Reduce Household Waste through product and packaging reuse

Design for reuse, repair and recycling

As well as buying too much stuff, we're buying products and packaging that can't be reused, repaired or recycled, because they haven't been designed for this. Whilst many of us may want to do the right thing when we no longer want the product or packaging, we are often prevented from doing so, because:

- producers are not designing and manufacturing products and packaging in ways that facilitate reuse, repair and recycling; and
- the infrastructure necessary to facilitate reuse, repair and recycling is not sufficiently developed to allow householders to adopt these methods of treatment of unwanted materials, even if they want to.

A major priority for producers has to be to design products and packaging that are suited to reuse, repair and recycling and to establish the infrastructure to support reuse, repair and recycling.

Let's start with packaging. Far, far too much packaging is single-use. It wraps and protects the product, but once we get that product home, the packaging is almost instantly discarded. I'll come to recycling in a moment, but first let's look at the potential for packaging reuse.

Direct packaging reuse

Just to remind you, direct packaging reuse is when packaging is used again for exactly the same purpose for which it was originally produced.

The best known examples (and possibly the only examples) of large-scale direct packaging reuse are currently limited to just two items: glass milk bottles and "bags-for-life". Yes, there are small-scale, local schemes involving glass wine bottle and beer growler refilling and reuse, but I'm struggling to think of any other examples of any scale. And reuse has to be done **at scale** if it is to make any meaningful difference. Local initiatives are good for raising awareness and saving very small amounts of materials, but unless we make changes that affect everyone and become the norm, we won't make the impact needed. So, we need producers and retailers to work together to make greater direct packaging reuse happen or, if not, then we have to accept that packaging reuse will have a relatively limited role to play in reducing Household Waste.

Apart from direct glass milk bottle reuse, in the past this also applied to beer and soft drinks bottles as well. These glass bottles often had deposits on them, to encourage consumers to bring them back for refilling. But in order to reuse a glass bottle maybe 20 times, it has to be strong, which means the glass has to be thicker and hence heavier. Over time, producers have reduced the thickness of the glass or replaced it with plastic or composite cartons, to save weight and material costs and to reduce the costs of transporting the packaged products. This is why reuseable glass bottles have gradually disappeared. Okay, the producers have reduced their financial, manufacturing and transport costs, but what has this cost the planet in environmental terms?

Glass milk bottles are starting to make something of a comeback as consumers look for ways to reduce their environmental impact and in particular their use of plastic. Local dairies are finding increasing numbers of customers want to have their milk delivered in glass bottles, rather than plastic jugs. But are there wider opportunities than just milk?

One of the reasons why glass milk bottles work so well is that everything is local and is controlled by one company, the dairy. The dairy delivers full milk bottles, collects the empties, washes and sterilises them and then refills them (from bulk collections from local farms) ready for the next delivery. It's a local, closed loop system.

However, once you move away from local supply, things get more complicated and the closed loop system is harder to achieve. Some dairies have diversified into things like fruit juice deliveries using the same glass bottles as milk. They buy in the fruit juice in bulk, bottle it and deliver it alongside the milk doorstep deliveries. This again cuts out the plastic or composite carton packaging normally used for fruit juices, so it's a good local, reuse system.

But could such a system work for other liquid products, such as carbonated (fizzy) soft drinks, beer and cider or wine? Clearly, for carbonated drinks you couldn't use milk bottles, because they wouldn't cope with the pressure and the lids are completely inappropriate. But could you introduce an alternative, standard form of glass bottle, with say a screw cap that could be used for beers, ciders and carbonated soft drinks? They could even be used for non-fizzy wine.

Such standard glass bottles used to exist to facilitate reuse. I can remember standard beer bottles that had a deposit on them. Consumers returned the bottles to get their deposit back and they were sent back to breweries for refilling (it didn't matter which brewery as the bottles were all the same (see "Deposit Return Schemes" on p. 150)).

There are pilot reuse/refill schemes being run by a very small number of the large grocery retailers, where the consumer purchases a reuseable glass bottle, has it filled with beer or wine and then brings it back for refilling once they have emptied and washed it. You have to remember to take your empty bottles with you when you go to the supermarket and the choice of beer and wine available for refilling is extremely limited.

One of my local, large supermarkets introduced a beer and wine refill scheme about two years ago. I dutifully bought the required reuseable glass wine bottles and beer growlers (refillable bottles called "growlers" typically hold three pints of beer), only to find the scheme discontinued after six months. Their explanation was "this was a pilot, prior to being rolled out across the country". I've seen no evidence of the suggested national roll-out and I and other well-meaning shoppers have been left with reuseable bottles and beer growlers that we cannot refill. Was this a pilot that didn't work? If so, why not? Or was this yet another example of a major retailer greenwashing their operations?

At the other end of the retailer store size, some enterprising "bottle shops" (shops that sell bottled beers) are selling draught beer in reuseable growlers. These shops offer a wider choice of attractive beers, but are dependent on the breweries being willing to sell them beer in bulk, rather than pre-bottled and the brewery loses the opportunity to promote the beer through on-bottle advertising. It's encouraging, but at the moment it's all very small-scale.

Maybe we should look at whether we could build on the growing interest in glass milk bottle doorstep deliveries and see if we could really boost glass bottle reuse and, importantly, without the need for a deposit-return system.

Glass and non-glass reuseable packaging is today starting to have a tiny toe-hold in grocery retailing. Consumers have been rightly appalled by the damage that is being done to our oceans by discarded plastics and are starting to challenge grocery retailers to provide plastic-free packaging alternatives. One way that the retailers are beginning to respond, and I mean "beginning" as this is really in its infancy, is to sell some food products loose (see below).

Deposit Return Schemes

In Scotland a Deposit Return Scheme was due to go live on 16 August 2023,[15] but has now been delayed until March 2024.[16] The scheme will cover: plastic bottles (with the exception of HDPE bottles, thus excluding most dairy products), metal cans and (originally) glass bottles. The delay has been caused by the UK Government insisting that glass bottles be excluded to bring Scotland into line with the rest of the UK. In Scotland, consumers will pay 20p per container, which will be refunded at collection points. The intention is to collect 90% of eligible containers by the second year of operation.

There are plans to introduce a Deposit Return Scheme in England, Wales and Northern Ireland from October 2025[17] (yet another long lead time). The scheme will cover plastic bottles and cans but not glass, and it appears the collected bottles will be recycled, not reused.

There has been much heated debate about whether a Deposit Return Scheme is needed or will work. Personally, I do not think it is needed nor do I think it will work. If we implement simple kerbside collection schemes, by which I mean co-mingled collection and MRF sorting, why do we need the proposed separate collection of just plastic bottles and metal drinks cans through a Deposit Return Scheme?

Think of the financial and environmental burden this will create. Consumers are already paying to have their recyclable Household Waste collected, so why make them pay again? Is 20p per container a big enough incentive to make people comply? I think not. And what about the environmental damage of manufacturing and then servicing all of the collection machines?

To me, a Deposit Return Scheme is unnecessary if standardised, simple, kerbside, co-mingled collection schemes are put in place.

Indirect packaging reuse

The most basic example of indirect packaging reuse is when consumers bring their own containers with them when they go shopping to temporarily package loose products, in order to take them home.

We are seeing the very early days of refillable packaging. But unlike glass milk bottles, where the milkman/woman takes the empty milk bottles away for refilling and leaves full ones in their place, the consumer actually fills their own container in the store (whether this is a specially purchased refillable glass wine bottle, the original plastic container for, say, liquid fabric conditioner, a paper bag provided by the store for loose tea or the customer's own natural fibre mesh bag for loose vegetables). Some of the very large supermarkets have introduced very effective weighing and labelling systems to make this happen.

But beware. The same supermarkets are also selling attractive, refillable containers to tempt consumers into buying even more packaging (albeit refillable) and apart from glass bottles (beer growlers and rescalable wine bottles), these are in very many cases made from plastic, thus adding to our plastic problems. You probably don't need to buy new containers to refill. You are likely to have suitable containers already at home, plastic sandwich boxes, used plastic take-away food containers or large ice-cream tubs can all be repurposed for the shopping trip. Really useful things you may not have are reuseable string or ventilated bags for transporting vegetables. But please don't buy plastic mesh bags (think of our oceans), buy natural fibre bags made from sustainable cotton or jute.

As I say, some of the large grocery supermarkets are introducing "trials" of selling some loose food (such as vegetables) and other household products (such as liquid washing cleaners). Some traditional small shops, such as greengrocers, have, however, been selling fruit and vegetables loose for as long as anyone can remember. Sadly, this is about the only example I can think of as bakeries, fishmongers and butchers generally sell their products in bags (unfortunately usually plastic bags), but could easily accommodate customers who took in their own bags or containers. But this just isn't happening.

Perhaps the biggest challenge to packaging reuse has been the shift of consumers to buying from large grocery supermarkets. These stores are attractive because consumers can buy everything they need under one roof, but the downside is that virtually everything comes pre-packaged (usually in plastic). Some leading supermarkets have run "trials" of selling loose foodstuffs, but these are limited in the range of products sold and are only in a handful of stores. When will such trials lead to a full roll-out across all stores so that loose selling becomes the norm? Progress is agonisingly slow for something so simple.

Of course, buying food loose does rely on the customer being willing and remembering to take their reuseable bags and containers with them and this requires a major change in customer behaviour. But the limited supermarket trials to date have shown both the infrastructure required is achievable with easy to use weighing and labelling systems being available. Now it just needs to be scaled up, so what's stopping it? My guess is that such "trials" are actually yet another form of greenwashing and the commitment from the supermarkets just isn't there.

And will consumers change; are they willing to move away from pre-packaged to products sold loose? I think, only if retailers convince them to do so, both through advertising and promotion, but ideally by making loose-sold foodstuffs cheaper than pre-packaged ones. Again, being realistic, it is economics that will drive change, not concern for the environment. Surely loose-sold products should be cheaper as the producers and retailers do not have the costs of pre-packaging the products? So here's a direct challenge to food producers and retailers – make loose-selling of suitable foodstuffs the norm and stop hiding behind "trials" and make such products cheaper than the pre-packed alternatives.

A second example of indirect packaging reuse is when we reuse unwanted packaging for new purposes within our homes. I discovered a very good example of this recently when I read that the inner bags from cereal packets (made from polypropylene) make very good freezer bags. All you have to do (once they are empty) is turn them inside out to rinse them and then use them as very tough storage bags. By doing so, you don't need to buy brand new freezer bags, thus reducing your new packaging consumption. This gives the bags a potentially very long second life.

Product reuse

Having talked about packaging reuse, what about reusing products that you or I no longer want? Just because I no longer have a use for a particular product, doesn't mean I should just throw it away.

If I no longer want something, there may well be other people who do. You can consider donating such items to charity shops, advertising it on websites such as Freecycle or, if valuable, you can think about selling it on eBay, Gumtree or sites like Vinted. Selling something you no longer want makes perfect sense and is now so easy to do. But if you're happy not to sell an item or it's too much hassle, then donating to charity is infinitely preferable to disposal and can help people less fortunate than you to buy perfectly good products for a very affordable price.

By and large, as consumers we've been programmed by advertising and society's norms to always want things we buy to be new. The mantra is: new is best, it's the latest version, it's clean and no-one else has touched it before. And whilst these points may be true for some things, particularly technology, which changes so fast that you probably do want the latest version, it's not true for all things.

I have a lot of hand tools and I've bought most of them second-hand, mainly because I really like the look, feel and function of old tools, but principally because I think many were better made in the past; the fact that they are still around demonstrates this. Plus, some are no longer available new.

Buying new or second-hand is about horses-for-courses; sometimes new is best, sometimes second-hand does the job, or is in fact better.

There is a trend today for "vintage" products, where the fact that the item is genuinely old (but still in good condition) has its own appeal. Examples are vintage clothes, vintage vinyl records, vintage furniture, artwork or old objects that are attractive in their own right and make nice display pieces.

There are many ways to buy used or second-hand products: eBay and Gumtree are obvious online sources (there are many others) and there are many antique or "junk" shops and charity shops on the High Street. Auctions can be a great source of second-hand or antique items and "For Sale" advertisements in shops or local newspapers often list second-hand goods. The popularity of television programmes like *Bargain Hunt*, *Flog It* and *Cash in the Attic* demonstrate the level of interest that exists in second-hand items and don't forget *Antiques Roadshow*.

Reuseable alternatives to replace single-use products

It is also possible and highly desirable to replace single-use products with reuseable products. A great example of this is switching from cling film to reuseable beeswax wrapping. Cling film is a single-use plastic product that is so hard to recycle that no-one attempts to and whilst it could in theory be recycled, it would be prohibitively expensive and if the principle of Producer or Consumer Responsibility were applied, the price of a roll of cling film would sky rocket.

A great alternative is beeswax wrappers and whilst these initially cost more than cling film, when the number of times they are reused is taken into account, plus the negative environmental impact of cling film, the economic balance tips in favour of beeswax wrapping. You can even make your own if you check out YouTube. Wouldn't it be great if television cooks and chefs championed this change, instead of wrapping everything in cling film. I say, "Cut the cling film".

So that's all about product and packaging reuse, now what about product repair?

Reduce Household Waste through repair

What do we do if our favourite product breaks or wears out? This is where we need to go back to a past way of thinking and rediscover the idea of repair.

But far too many products, from toasters to mobile phones, are simply not designed for repair and, even if they are, where is the easy-to-access infrastructure that allows consumers to take their broken or worn products for repair? Again, I'm struggling to think of examples where this does exist, but apart from shoe repair (which is encouragingly very common) and to a lesser extent clock and watch repair and clothing repair, what else is there? If we're going to make repair meaningful, it has to be as easy as having your shoes mended.

Repairing something is a form of "waste" reduction. Repairing returns what was seen as "waste" into a fully useable, fully functioning whatever-it-was-originally. So, repairing an Xbox or PlayStation, a mobile phone or a shoe, a handbag or an item of clothing, gives us back what was once thought of as "waste". This isn't recycling, because we don't change the product, we just fix it, so it's waste reduction.

There are two basic ways of repairing a broken or damaged product. You can do it yourself or you can take it to a repair shop and pay to have the work done.

I was brought up in a very practical household and we didn't have much money, so if anything got broken or worn out, we always tried to fix it. It is very satisfying to repair something and bring it back into use, particularly if you do it well and with pride. But it is a sad fact of life that too many people today don't have the skills, tools or inclination to carry out practical repairs. As a society we seem to have lost this ability, which I think is very sad. The only ray of hope is that if you do want to have a go at repairing something there is probably one or more videos on how to do it on YouTube.

But, whilst one of the key ways of reducing Household Waste is to repair products so they can continue to be useful, something that makes absolute environmental and financial sense, producers are not keen on us doing this. In fact, they don't like it, because if you repair a product, you don't need to buy a new one and our consumer society is based on always buying a new one; we consume, we don't preserve. So this is another attitude we need to change, both in the minds of consumers and of producers. For producers this represents a new business opportunity; okay, if successful they would sell fewer new products, but they could develop a whole new side to their business offering repair services.

Increasing product repair

For product repair to be possible, we need products to be designed and manufactured in ways that facilitate repair and we need the infrastructure to be developed to allow consumers to adopt this method of treatment.

Producers should design products in such a way that they can easily be repaired and not simply design them to be as cheap as possible to manufacture. This will require a major change in the thinking of producers, but it is not rocket science, it's more a case of good design practice. Yes, it is very likely to add to the cost of manufacturing the product and this will be passed onto the consumer as an increased price, but in all probability the cost of the product to the consumer over its extended lifetime will be reduced. And this is without taking into account the environmental benefits of repairing, rather than disposing of, a product.

Repairing something doesn't have to be the craft form that we have all enjoyed seeing on the television programme *The Repair Shop* where very skilled craftspeople bring damaged and worn objects back to life. Yes, this approach is extremely worthwhile, particularly for much loved or valuable items, but often repair is simply a question of sourcing a single broken component and fitting it. And the best place to go for replacement parts, apart from the internet generally, is eBay.

You will be amazed at just how many replacement spare parts are available and at little cost. Just type in the name of the product followed by "spare parts" and up pop multiple vendors. By replacing what is often a small and simple item, a whole product can be repaired, either for your own use or for resale to make some money. Some people make a living out of rescuing broken items from junk shops or car boot sales, repairing them at little cost and selling them on for a profit. This demonstrates how something seen as "waste" can still have an economic value, as well as preserving the product along the way and so reducing "waste".

As an aside, there is a Japanese art form called Sashiko which makes a feature of clothing repairs. Originally Sashiko was the technique that used inexpensive white thread to strengthen and repair everyday clothes made from cheap indigo-dyed cloth. But people gradually recognised the artistic value in these repairs and the techniques of Sashiko and Boro (when a garment has been repaired many times using Sashiko stitching), so much so that they have now become a way of making previously unwanted clothes not only repairable, but in a way that makes them more attractive. We always used to try to hide repairs, for example by "invisible mending" of clothes, and in some instances this is still preferable, but here is an alternative, where this is not possible, that celebrates the repair and makes it a feature.

But going back to product repair, I recognise that producers want to make products as cheaply as possible, so they design products for the cheapest methods of manufacture, not for repair. They glue or fuse things together (I'm thinking plastic here) rather than making parts separable for repair, for example by using a screw to join components. And all because it means the product is cheaper to make and when it fails we have to buy a new one and throw the old one away. This is the worst kind of linear consumption in action. But it doesn't have to be like this. Producers could create a whole new side to their businesses based on repair, whether simply making replacement parts available or actually providing a repair service.

The other challenge to the practice of repair is the cost of repair. Many products that we buy are manufactured in countries that have lower labour costs than in the UK, which is why they are cheaper to make than if made in the UK. But if these products are to be repaired, this would in all likelihood happen in the UK, with our higher labour costs, making it relatively expensive to repair compared with buying a new, cheaper replacement product. So what do we do? This is where I think true Producer Responsibility lies.

In order to boost the levels of repairability, the first thing that needs to happen is for producers (manufacturers of durable products, clothing, household items such as furniture, etc.) to change how they design and make products. The key changes needed for producers to reduce Household Waste are by:

- making products that last and eliminating the contemptible practice of built-in obsolescence; and
- designing products that can be repaired or upgraded and putting in place the infrastructure required to facilitate such repair or upgrading.

As I said in Chapter 6, if we are to move to a Circular Supply Chain for products, the first thing we need to do is to implement changes to the Supply Chain, to collect unwanted products and to create a repair industry.

The concept of a much expanded repair industry opens up a number of interesting and I think quite exciting possibilities:

- first there is the idea of local, generalised repair shops, that can repair any given product or, if they can't do it themselves, send it back to the original producer for repair (some such shops already exist, but there are very few); wouldn't this be an amazing business and job creation opportunity?
- producers themselves could offer a new service of repair, either as a direct service to the owner of the product or by introducing take-back schemes where the consumer trades in their old product when buying a new one

and the old product is actually repaired and resold, thus creating new markets for refurbished products.

Thus, product repair, as well as being good for the planet, could create new business opportunities for producers.

Reducing Household Waste – summary

To **reduce** Household Waste we need to:

* reduce our consumption;
* design products and packaging so they are suitable for reuse, repair and recycling;
* increase packaging reuse;
* increase indirect product reuse; and
* rediscover the lost art of product repair.

And I'm sorry if I keep banging on about this, but if such "waste" reduction is to make any real impact, it has to become the norm and be carried out **at scale**.

But given there will always be stuff we can't reuse or repair or basically that we no longer want, the next best option is to recycle it.

Recycling of packaging

I'm going to talk about the recycling of packaging and products separately. Let's take packaging first.

A new definition of "recyclable"

If you remember, in Chapter 3, I introduced a new definition of the term "recyclable" to mean the following and **only** the following – recyclable materials are only those that:

* are collected at scale:
    * for products this means take-back schemes must exist for the unwanted product via retailers (see Chapter 6); and
    * for packaging, the materials must be collected by all Waste Collection Authorities in their kerbside recycling collection schemes;
* can be separated into individual materials:
    * for products, a dismantling infrastructure must exist for the particular product; and

- for packaging, this must be capable of being sorted into individual materials in a MRF; and

- are reprocessed to become the equivalent of virgin raw materials, economically and at scale.

This is what recycling really means and we need to stop producers from misleading consumers by labelling their products or packaging as recyclable, when in reality they're not.

I refer you to my rant on packaging labelling in Chapter 8 and to my comments in Chapter 7 where I quote Tom Szaky who says (and I'm paraphrasing here) that all "waste" is recyclable, but at a cost, but that cost can be prohibitively high.

But it's not just about financial cost, it's also about practicalities. I've already said that recycling has to be kept simple and carried out at scale, if it is to make a meaningful difference. So, that means WCA co-mingled kerbside collections and MRF sorting of Dry Recyclables (packaging). I'll come to the recycling of products in a moment.

The only way to achieve the levels of recycling of packaging needed is through kerbside collection, not through HWRCs or bring collection or plastic packaging being posted back to a reprocessor. Forget individual material drop-off collections or reverse retailing where consumers take their unwanted materials back to the store they bought them from (this is the idea behind reverse vending machines that give consumers their bottle deposits back); this just isn't going to work. Follow the KISS principle and then make producers only use materials that are truly recyclable.

If we accept my proposed definition of recyclable materials, the labelling of packaging becomes oh-so-simple. We only need one symbol for all recyclable packaging and I've proposed two possibilities in Figure 10.1 (if the first variation was adopted we will finally have a use for the Mobius Loop symbol). I've included the words "RECYCLE AT HOME IN KERBSIDE COLLECTION" just to make it completely clear what householders should do with the packaging. But in theory, all the OPRL needs to say is "RECYCLE", but maybe it would be better to spell it out.

These two OPRLs give a clear and simple definition of what is meant by "recyclable" and it, or something similar, needs to be universally accepted by all producers and retailers and explained simply to all consumers and householders. Only in this way can consumers make informed choices about what they buy and householders be expected to know what to do with different unwanted materials.

Producers in particular should only label a product or packaging as recyclable if it complies with this definition and WCAs should communicate

FIGURE 10.1   My proposed truly recyclable OPRL.

very clearly to householders exactly what will and will not be accepted in the Dry Recyclables kerbside collection arrangements.

Too many producers currently cop out of their responsibility to ensure the materials from which they make their products and/or packaging are actually recyclable, as I have defined above. They label them as "recyclable" when what they really mean is "recyclable in theory" (see Chapter 8 for more discussion on this).

Collection and sorting

For the recycling of Household Waste to happen on a much bigger scale than at present, we need three changes to happen:

* producers to "design with the end in mind" that is, to only use those materials that are recyclable as defined above and can therefore be handled by the Household Waste kerbside collection and MRF sorting recycling infrastructure;
* WCAs and WDAs to provide appropriate and simple collection arrangements (in reality meaning co-mingled kerbside collection – see later in this chapter) and to contract for the necessary treatment infrastructures, including local MRFs and accessible reprocessing capacity; and
* householders/consumers to actively and correctly separate their unwanted materials in a much simplified manner (KISS).

Recycling collection has to be done at scale and kept simple

I want us to be realistic about recycling. Recycling is not the panacea that many people think it is. This is because very few materials are infinitely recyclable and because our current arrangements for unwanted materials collection are inadequate.

Yes, we have come a very long way in terms of implementing kerbside recycling, but we need to do much more if we are to achieve anything approaching a circular pattern of consumption.

Reported recycling rates of 45% in the UK[18] make it sound as if we're doing really well, but we're not. We should be aiming for much, much higher levels of recycling. We need to collect more materials and in greater quantities; too much of our recyclable materials are still being sent for energy recovery or disposal, because:

- WCAs don't collect all the potentially recyclable materials;
- WCAs can't collect some materials, because they are not designed to be recyclable (such as composite drinks cartons and products and packaging made from hard-to-separate materials);
- householders are confused by the collection arrangements and don't separate out all the materials they could;
- some householders just don't care (let's be honest) and so don't separate their Household Waste for recycling collection or do so badly, potentially contaminating materials that are collected for recycling;
- there is insufficient reprocessing capacity, in the right places in the UK, to handle all of the materials that could be collected; and
- the end markets for some collected materials are not paying sufficiently high prices to make the collection of some materials financially worthwhile.

Let's start at the beginning by looking at the collection of Dry Recyclables from households. If we are to increase collection rates to what they need to be, collection from the householder needs to be operated at scale. By this I mean every householder correctly separating and setting out their recyclables for kerbside collection. For this to happen the collection arrangements need to be simple and convenient for the householder to use. I'll say this again, **collect at scale and make it simple to use**.

KISS (Keep it Simple Stupid)

If collection rates are to increase we need the collection arrangements, and by this I mean the collection containers and the degree of sorting required from the householder, to be really simple. We will only drive up the recycling of Household Waste to the levels that I think are achievable, if we make the collection arrangements as simple and easy for householders to use as throwing something away is now. This needs to be our guiding principle, so I'll say it again: **make collection for recycling as simple as throwing something away is now**.

I propose that every household should have only one container for Dry Recyclables, a wheeled bin, with all the glass, paper, plastic, cans, etc. mixed up; just chuck it in – easy-peasy. This is called co-mingled collection. Only if it's easy will there be a chance that everyone (or the majority) will do it and do it correctly.

Will people do it correctly? Will householders put the right items in the recyclables bin, unwanted food in the food "waste" bin and the small amount of what's left in the residue bin? No, not all will, people will make mistakes and some people just won't care. But by a combination of education and having a single, standard UK-wide system, with the same materials collected for recycling across the country, surely we can get people to adopt the right habits and we want it to become a habit so people do it and do it right, without thinking.

There was an article in *The Times* recently, picked up from the *i* newspaper, that Defra was planning to publish guidance in the summer of 2023, to discourage householders from wishcycling, which is putting items into their recycling container hoping they will be recycled, but which in reality won't be,[19] Apparently, wishcycling not only increases the amount of Household Waste rejected when the collected materials are sorted, but can cause significant contamination, for example if greasy pizza boxes are included in cardboard collection. It's important to encourage householders to put only the right materials into the recycling collection container, but this is best achieved, as I say, by having a single, standard UK-wide system that is easy to use, i.e. co-mingled collection.

I know this is a radical proposal, but I absolutely believe that this is the only way we will make recycling collection the new norm. Some people, including some material reprocessors, argue that materials such as collected used paper can only be of a sufficiently high quality if it is separated from everything else, by the householder, at the point of collection.

The Welsh Assembly Government argues that "To achieve closed loop recycling (especially for paper, glass and textiles) the right collection system needs to be in place so that the recyclate is not contaminated. For household waste this is best achieved through kerbside sorting whereby recyclable materials are sorted at the kerbside into different containers or compartments on a collection vehicle".[20] But I refer you back to my comments under the heading "Kerbside collection" in Chapter 4, where I described the practical inefficiencies of this approach.

When I operated a MRF back in the 1980s, we collected co-mingled paper, cardboard, metal cans and foil, plastic bottles, and glass bottles and jars all in the same wheeled bin which was emptied into a compactor collection vehicle and tipped onto a concrete floor before sorting. We had very little glass breakage, because the other materials effectively cushioned and protected the glass.

My argument is that asking householders to separate different packaging materials into a number of different containers is too complex and involves too many collection containers per house, so that a very large percentage of householders just won't do it or don't do it correctly. I advocate going for the simplest solution, make recycling as easy as throwing stuff away is now: one bin for all recyclables, job done.

Given that the unwanted materials will be collected as co-mingled materials, each WCA/WDA will have to contract with a local MRF to carry out the sorting and separation of the collected materials and this in turn will require many MRFs to be built.

My proposed simplified Household Waste collection system

I think every household should have **only three** basic collection containers:

- the largest container for co-mingled Dry Recyclables, say, a 240-litre wheeled bin (this is the standard size of the vast majority of wheeled bins currently used for Household Waste collection for disposal);
- a smaller container for residual "waste", say, a 120-litre wheeled bin (it should be smaller than the Dry Recyclables wheeled bin, both because it just doesn't need to be as big as is now the case and because it will send the message that recycling is the priority, not disposal); and
- the smallest, a food "waste" container, such as the 23-litre caddy used by WCAs who operate food "waste" collection schemes now.

One of the benefits of this approach is that virtually every household already has a 240-litre wheeled bin, so these just need to be repurposed to become THE recycling bin (each WCA may need to organise a mobile washing service to clean these bins before the change of use). The additional bin cost is thus limited to the smaller 120-litre wheeled bin and the food "waste" caddy, where these are not already in place.

There should also be an optional garden "waste" wheeled bin for separately collecting green "waste". But separate garden "waste" containers should be paid for by the householder (as most WCAs require now), as many householders either do not have a garden or, if they do, carry out home composting. It would be inequitable for these householders to pay for the costs of garden "waste" collection through their Council Tax.

Such a collection arrangement would keep things simple for the householder, thus increasing the chances of high levels of participation. If this could be adopted as a standard approach by all WCAs, it would have the added benefit that when people go on holiday in the UK, they would already know how the system operates, thus boosting recycling collection volumes in tourist areas (where they are notoriously low). If we have a standard system of recyclable collection, then we can really start to change consumer attitudes; it becomes the norm and people can do the right thing without having to think about it.

I also think we should phase out all bring collection sites, except at HWRCs where these are **only for materials not collected via kerbside collection**. Again this is a radical idea, but bring collection only accounts for 1% of all collected recyclables[21] and it's financially and environmentally uneconomic (as explained in Chapter 4 under the heading "Drop-off").

I honestly think we have to make the kerbside collection of Dry Recyclables, **the** method of Household Waste recyclables' collection; forget bring collection, including at HWRCs, for Dry Recyclables. Have one system, explain it to householders and make it standard across the UK.

Collection frequency

It used to be that everyone's rubbish was collected once a week. A few years ago, actually quite a few years ago, WCAs started to introduce fortnightly Household Waste collections, primarily to reduce costs. There was perhaps an understandable public outcry. But despite the initial misgivings, fortnightly collections have become not only commonplace, but the norm, and there have been no real adverse effects as a result.

One of the key reasons fortnightly collection of Household Waste has been successful is because many WCAs have implemented separate weekly food "waste" collections. This has taken the nasties out of the residual Household Waste bin. So all the arguments about the life cycle of the common housefly being less than a fortnight have gone out of the window. Things that can rot, smell and attract flies are still being collected once a week, but in a very different and much more sustainable way.

How often should Household Waste be collected? I believe that three of the collection containers, that is, the co-mingled Dry Recyclables, the residual "waste" and the garden "waste" wheeled bins should be collected on a fortnightly basis to maximise collection vehicle efficiency. The contents of the smaller food "waste" container should be collected on a weekly basis (because of health concerns).

What is so telling is that once you separate food "waste" from all the other discarded materials, the cleanliness and quality of these other "waste" materials goes up significantly.

By collecting co-mingled Dry Recyclables one week and residual "waste" the next, the same fleet of collection vehicles can be used (known as compactors as they squash the load to maximise the amount of material collected in the vehicle). This minimises the cost of the collection vehicle fleet.

The materials that are actually recyclable

Which materials should be included in a co-mingled Dry Recyclables' kerbside collection?

As I said in Chapter 3 (under the heading "Dry Recyclables"), when talking about Household Waste recycling schemes there is a minimum range of what are called Dry Recyclable materials that are collected. These are:

- paper;
- cardboard;
- glass containers (bottles and jars);
- metal food and drinks cans (steel and aluminium);
- aluminium foil;
- plastic bottles; and
- some hard plastic containers, i.e. food packaging, such as tubs and trays.

These are the materials that are normally collected in a kerbside recycling collection scheme.

These materials all therefore fall under my definition of recyclable and should be included in future co-mingled collections. But there is one important addition I would like to add, that of plastic film products (also called "flexible plastics").

Remember, almost all kerbside schemes exclude plastic film. From Table 2.2 (Chapter 2), plastic film constitutes 3% of Household Waste, amounting to nearly 900,000 tonnes each year in the UK. If all of this plastic film was recycled, this would increase the plastics recycling rate by 50%. However, there are two problems that need to be overcome, before this can happen.

The first is the sheer number of different types of plastics being used, both dense (rigid) plastic like bottles, trays and pots, and plastic film. Separating all of these plastics into individual polymers, once they are no longer wanted, is not a realistic option, plus not all of the polymers are readily reprocessed.

The solution to the problem of plastics recycling is, I suggest, really rather simple. Producers should limit the number of polymers used in packaging to only those that are readily recyclable, which means Polyethylene Terephthalate (PET) and Polyethylene (PE) as these are currently the most commonly used polymers and the reprocessing techniques for PET and PE are proven and reprocessing plants exist, but not in sufficient numbers and not in all the right places. So we need the plastics reprocessing industry to invest in additional plants but, critically, to site them in the right places to be as local as possible to the sources of Household Waste. But crucially, we need producers to cut the long tail of minority-use polymers and to only use PET and PE.

Could the number of polymers in use be reduced to just two? It could be if the Government legislated, but I suspect the Government would be unwilling to do so, as it is usually reluctant to do anything that could potentially harm trade or restrict market economics, so what would convince producers to make such a change? Well, I can think of three things:

- consumer pressure – if we stop buying products and packaging made from other polymers and tell producers this is what we are doing, then they will listen (but can enough consumers identify different polymers and, let's be honest, do they really care enough to act?);
- retailer pressure, if retailers see consumer resistance to the extent that it is affecting sales, then they will lean on producers to change; and
- maybe even producers will see the environmental sense of this change and (cynic that I am) exploit the green image they could create for themselves by reducing the number of polymers they use.

But this will only happen if the other pieces of the jigsaw are put in place: the kerbside collection, the MRF sorting and the building of additional plastics reprocessing capacity. It all has to happen and in a co-ordinated manner.

The second problem with recycling plastic film is that it is difficult to separate plastic film in an automated MRF, because the film behaves physically like paper and so it is difficult to separate out. Until this can be done, WCAs will continue to exclude plastic film from kerbside collections. As I said in Chapter 9, currently the only place you can take your plastic film for recycling collection is a relatively small number of large supermarket stores. The amount collected is, I suggest, miniscule and this type of collection is inefficient, both financially and environmentally. It's greenwashing as this is not a real solution to the problem of plastic film.

The solution to plastic film recycling is not so straightforward but, as an engineer, I would be amazed if MRF designers could not come up with a solution. Just off the top of my head, if you could create a stream of mixed paper and plastic film (not so hard to do), perhaps you could separate them using water, since paper absorbs water and becomes heavy but plastic does not – just a thought.

There is one glimmer of light for plastic film collection and reprocessing. The FPF FlexCollect project[22] is being run by a consortium comprising plastic "waste" research groups, Defra, waste management companies, producers who use plastic film and other stakeholders. It's a three-year project designed to work with a small number of local authorities to develop a better understanding of plastic film collection options, with the intent of launching guidelines in 2027 to enable local authorities to include plastic film collections in their kerbside schemes. Encouraging, but why will this take three years? And this is just to prepare the guidelines; any new collection arrangements will take further time to implement and new reprocessing capacity will need to be built if such collections arrangements are successful. Come on guys, we need greater urgency here.

I suggest that the simple, pragmatic solution to the problem of plastic film is to require only polyethylene film to be used in packaging and then to ask householders to include all plastic film in their recyclables bin.

Given it will take time to introduce such a change, in the interim, we should require all producers to very clearly mark their polyethylene film packaging as being made from polyethylene and asking householders to only put polyethylene film in their recycling bin, but maybe this too complicated.

And don't forget that I have argued that so-called compostable or biodegradable plastic film is problematic and should not be used.

Packaging recycling and organic recovery

I've summarised the above more circular approach to the reuse and recycling of packaging, plus the collection and treatment of organic materials, facilitated by a simplified kerbside collection system. This is illustrated in Figure 10.2.

This book is aimed at **all** the players in this newly defined Circular Supply Chain. Figure 10.2 shows the apparent complexity of the Circular Supply Chain and the number of players involved. I say apparent complexity, because

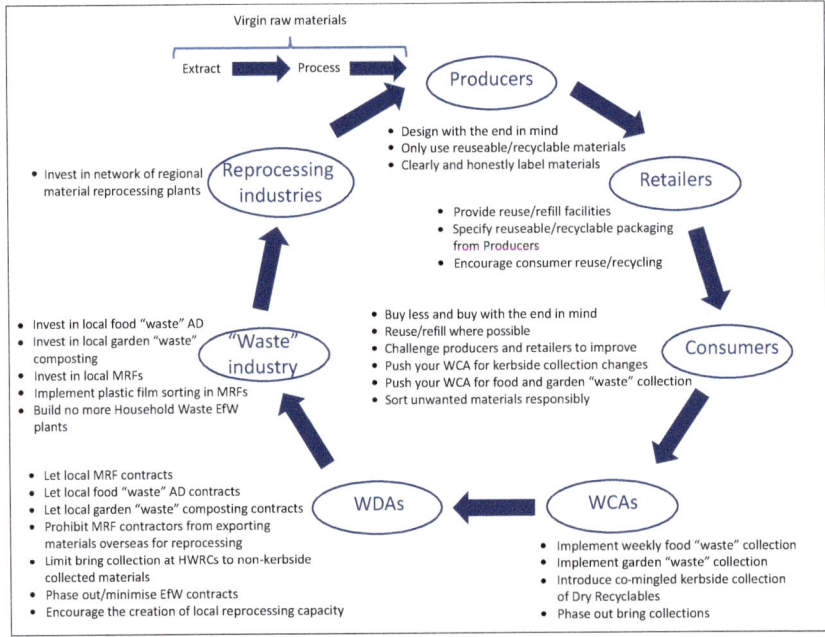

FIGURE 10.2 The Circular Supply Chain for Dry Recyclables and organic materials.

there is a lot going on and many players are involved, but if we break down this cycle, it is actually a series of relatively simple and discrete steps carried out at each stage by only one or two players; it's not so complex if we take it step-by-step.

What is different and very important is that each stage in this supply chain needs to take into account the subsequent steps. For example, consumer products and packaging need to be designed by producers at the very start of the cycle, in such a way as to facilitate reuse, repair and recycling, rather than the subsequent stages trying to figure out what to do with products and packaging that have not been so designed.

This is why I said earlier in this chapter that we've been approaching the development of recycling from the wrong direction. All our efforts to date have been led by WCAs and WDAs who have been doing their best to recycle **existing** products and packaging, instead of products and packaging being redesigned to be recyclable. **This is a fundamental point**.

Unless products and packaging, are designed to be recyclable, the WCAs, WDAs and "Waste" Management Industry will always be operating at a suboptimal level, trying to make the best of what is currently manufactured, rather than the products and packaging being optimised for recycling. So here we have a key change that is required and it's simple; producers need to design-with-the-end-in-mind. And as an engineer I say that this is not so hard to do. Yes, it may increase some product or packaging costs, **but it may also reduce some**. What is needed is better design and I think that's something that this country is pretty good at.

Yes, Figure 10.2 shows there are a lot of players involved, but it all seems relatively straightforward doesn't it? Well actually no, not until we apply the golden key to unlock this cycle, which is **designing** the products and packaging for reuse, repair or recycling. Blindingly obvious isn't it? Then the cycle becomes straightforward.

Some producers are starting to design products and packaging that are more suited to such a recycling infrastructure, but I have to say that they are in the minority and progress is painfully slow. Plus, the need for much more sorting and reprocessing capacity just isn't being addressed.

We need multiple players to come together to make these changes happen, but we should start with just two:

- we need the Government to define what is meant by "recyclable"; and
- we need producers to design products and packaging to be "recyclable".

Obviously, we then need WCAs to implement collection arrangements that deliver the new definition of "recyclable" packaging for recycling and organics for recovery.

But there is also a lot that you and I can do to drive this change. We are the consumers and if we change how and what we choose to consume, then the producers and retailers will notice and start to change what they do; that's the way the consumer world works.

Summary of recycling changes required

To summarise the above, we need:

- a new definition of "recyclable";
- for this definition to be applied to packaging and for producers to stop greenwashing their packaging;
- producers to design-with-the-end-in-mind;
- to make kerbside, co-mingled collection the default recycling collection method for Household Waste (KISS) and phase out bring or drop-off collections; and
- to limit kerbside, co-mingled collection to a standard list of materials, including plastic film.

The recycling of products

As I said in Chapter 6, we need a different approach to the collection of unwanted products by implementing changes to the Supply Chain that will allow and encourage consumers to take unwanted products to retailers, rather than using kerbside collection. These retailers will then partially reverse the Supply Chain in order to feed these products into an expanded repair industry or if unrepairable, into a dismantling industry that generates individual materials for reprocessing. This latter route will facilitate the recycling of materials from unwanted products, because unless products are dismantled, the materials within them cannot be recycled.

Should we include WEEE in co-mingled collection?

One category of products that some WCAs include within their kerbside collections is small items of WEEE, such as mobile phones, radios and personal stereos. However, these are only in box-based schemes where the box contents are hand-sorted and separated into a separate compartment within the collection vehicle.

I am, however, advocating co-mingled kerbside collection of Dry Recyclables in one wheeled bin, which would be collected in a standard compactor collection vehicle, which involves the collected materials being compacted (squashed). There is an industry concern that if small WEEE was included within such collection vehicles, the compaction could cause damage to the WEEE items and could risk any lithium-ion batteries, contained within the WEEE items, to be crushed, leading to the risk of a fire.[23,24] For this reason, I have reluctantly concluded that WEEE items should not be included in the kerbside, co-mingled collection and that they should be collected using the options outlined in Chapter 6.

Recovery

First of all it is important to distinguish between what we mean by recycling and recovery.

Recycling is the collection and processing of a particular material to produce a clean, reuseable form of that material that can be made into new products. For example, recycling glass from discarded bottles into new bottles or unwanted aluminium drinks cans into new aluminium drinks cans.

In contrast, recovery is the collection and processing of materials to recover either an alternative material to the original one or energy and this can happen only once; this is not a circular method of treating materials.

There are currently four ways to recover what can't be recycled:

- down-cycling materials into a new material or product that is of a lower quality than the original materials;
- anaerobic digestion of food "waste";
- composting of garden "waste"; and
- EfW incineration of mixed materials to recover energy.

I've talked about down-cycling in Chapter 5, so let's take the next three in turn.

Recovery of organic materials

Anaerobic digestion of food "waste"

Every year, we each buy and consume significant quantities of food, but we also throw a huge amount of it away. In 2018 we consumers collectively bought 43 million tonnes of food and drink.[25] Of this we threw away 6.6 million tonnes (15%) in our Household Waste, that was either food we couldn't actually eat (like the inedible bits of a chicken carcass, vegetable peelings or meal left-overs on someone's plate) or food that was past the use-by date printed on its packaging. The organisation, WRAP, estimated that 4.5 million tonnes (70%) of this "waste" was food that we could have eaten (that is, excluding the inedible bits).[26]

But this is only the tip of the food "waste" iceberg. There is a universally accepted statistic that one third of all food produced globally is wasted. This is appalling. Let me quote you a passage from a recent book *Earthshot: How to Save Our Planet*:

> Our top priority should be putting an end to food waste. … It is a scandal that a third of all food – grown at such great expense to our forests and grasslands, rivers and oceans and the atmosphere … is simply thrown away. It works out at roughly 100 kilograms wasted by average citizens of developed nations every year. … Not all of this is done personally. Much of it disappears into bins at the back of supermarkets, restaurants, canteens and warehouses. … WRAP found that almost a fifth of lettuces grown in Britain never get beyond the farm gate, along with a tenth of strawberries. But whoever is responsible, it is a huge waste of all the land and water, energy and farm chemicals and time required to produce it in the first place.[27]

And this is when humankind is struggling to feed all the people on the planet, land for food production is increasingly scarce, rain forests are being felled to create more land for food production, agricultural chemicals are polluting our waterways … I could go on. But as consumers, or even as a country, we can't address all of the global problems of food "waste", but we could do so much more than we currently do.

The UK figure of 6.6 million tonnes of food "waste" is just the food "waste" that arises within households (a huge 24% of Household Waste).[28] Food "waste" also arises in other sectors as follows: 0.8 million tonnes in the food manufacturing sector; 0.3 million tonnes in the retail sector; and 0.8 million tonnes in the hospitality and food service sector (restaurants, pubs, events venues, street food vendors, etc.) This suggests a total UK food "waste" figure of 9.5 million tonnes per annum.[29]

Coming back to Household Waste, WRAP also estimated that of the 24% food "waste" within Household Waste: 20% was anaerobically digested or composted; 45% was recovered either through EfW incineration or spreading on land; and 35% was disposed of either to landfill or the sewers (and therefore ultimately to landfill or, appallingly, sometimes into our rivers and seas).

That only 20% of household food "waste" is being recovered through anaerobic digestion is shocking and shows how the kerbside collection of food "waste", coupled with anaerobic digestion, could make a massive reduction in the amount of food "waste" currently being recovered as energy through EfW incineration or simply disposed of to landfill (currently a combined total of 80%).

WRAP estimates the figure of 6.6 million tonnes of household food "waste" would fill 66,000 three-bedroom terraced houses, equivalent to the population of a town the size of Peterborough. Apart from being a disgusting image, this is an appalling mismanagement of the food products and everything that goes into their production, when there are tens of thousands of people in the UK struggling to get enough to eat.

There is apparently some good news though. WRAP states that between 2007 and 2018 the amount of household food "waste" that could have been eaten reduced by 1.6 million tonnes (26%), representing a saving to householders of £4.8bn per year. They give an interesting analogy in that this equates to a reduction in CO_2 of 2.4 million cars off the road.

As householders we are currently throwing away 6.6 million tonnes of food each year, either food we just don't want or food that is past the best-before date or use-by date printed on its packaging.

Let's take a moment to look at these dates. There are often three dates printed on food and drink packaging: the best-before date; the use-by date; and the display-until date. I'll let WRAP explain what these mean:

> The term "Best Before" … indicates the period for which a food can reasonably be expected to retain its optimal condition (e.g. bread will not be stale) … Providing food is stored in appropriate conditions and has not become otherwise contaminated … it will be safe to consume for a period of time following the expiry of a "Best Before" date, but it may not be at its best.
>
> Consumers should not eat food which has an expired "Use-By" date unless it has undergone an additional process that makes the food safe prior to the date of expiry. Additional processes can include cooking and freezing.
>
> "Display Until" and similar dates are usually for the retailer's stock control purposes and are not there to provide information on food safety or food quality. There is no legal basis for a food to carry these types of dates and WRAP research has shown that these can be confusing for consumers and has resulted in good food being thrown away.[30]

Whilst as householders we will always want to get rid of what I would call real food "waste" (the chicken carcasses, etc.), I think we should treat food best-before dates with caution and a degree of scepticism. Best-before dates are a guide and whilst you should never take risks, food is often perfectly edible after the stated best-before date. Your eyes, nose and brain will tell you if something is really out of date (if it's weeks past it's best-before date don't eat it!); you just need to be sensible and careful, but a lot of food that is past its best-before date is perfectly edible and safe to eat. And, of course, by not throwing perfectly edible food away, you could save yourself some money.

If you do have to throw away food that is past its best-before or use-by date (and, as I say, please don't take risks) and it is still in its packaging, if you have a food "waste" collection service, take the time to separate the food from the packaging and set out separately – the food for recovery and packaging for recycling. This way both the food "waste" can be recovered and the packaging can be recycled.

It was interesting that during the 2020 Coronavirus pandemic, many people thought more about using what food they had, rather than throwing it away, because during the lock-downs, food was at times difficult to obtain. For some people this highlighted just how much food we waste and how much could be saved.

There is a huge potential to increase organic recovery, particularly with regard to food "waste". The current recovery rate for food "waste" is a paltry 20%. This is for two reasons:

- not all WCAs provide a food "waste" collection service; and
- for some WCAs that do provide such a service, the take-up of the service by householders can be disappointingly low (one WCA I spoke to told me that they have only been able to persuade 50% of householders to participate).

But collecting food "waste" separately has three real benefits:

- it allows this type of organic "waste" to be recovered as useful digestate (a soil conditioner) and combustible gas that can be used for energy generation;
- it significantly reduces the weight of Household Waste sent for disposal (Household Waste is measured by weight not volume, and given the high water content in food "waste" it is a heavy contributor to Household Waste); and
- it removes the often unpleasant, yukky mess that food "waste" can bring, thus avoiding the contamination of other materials.

Taking this last point, one of the things that is interesting about a WCA implementing a separate food "waste" collection service is the impact this has on the residual Household Waste for disposal. If you separate items for recycling and then set out food "waste" and garden "waste" for separate collection, what is left is not only surprisingly small in quantity and weight, but it is also surprisingly clean and easy to manage.

My disposal wheeled bin has very little in it after I have separated out the Dry Recyclables, food "waste" and garden "waste". It's mostly unrecyclable packaging, such as plastic film, composite materials that I can't separate such as juice cartons and the odd contaminated item, such as a used polishing cloth or things I have used to wipe up spills. My Household Waste for disposal is collected every fortnight, but as I've said it is not unusual for me to only put this wheeled bin out for collection once a month, which if we all did this would have major, positive implications for WCA collection operations and costs.

As you will see in a moment, the scope for increasing the recovery of food "waste" is huge, from the disappointing 20% now to what? Why not to 100%?

There are thus two changes needed to facilitate the potential improvements in food "waste" recovery. First, we need all WCAs to provide a small food "waste" caddy that is collected weekly. As many, many householders have already found, this is not a real imposition and you quickly get used to separating food "waste" from other unwanted materials; it's even quite satisfying.

Second, we need our local WDA to contract out the processing of the collected food "waste" in a suitable anaerobic digestor. These do not need to be large in scale and are commonplace. Many farmers use anaerobic digestion to process their animal wastes, so there are many contractors who are willing and able to provide the required service (at a price of course).

The good news is that the Government in England planned for all WCAs to provide food "waste" collection services by 2023. Well, that date has passed, so let's hope this happens.

Composting of garden "waste"

Our performance with regard to garden "waste" is much better than for food "waste". Currently, 90% of all collected garden "waste" is recovered either via kerbside collection or HWRCs. This demonstrates that householders are willing to separate garden "waste" for commercial composting. And a lot of householders compost their own garden "waste" at home, so this material doesn't show up in the collection figures.

Home composting

As I explained in Chapter 5, many people have a home compost bin in which they recover garden "waste" and uncooked food "waste" as compost. They use the resulting compost as a valuable addition to their garden, adding humus and nutrients to their garden soil.

And whilst this is a long-standing and excellent way to treat garden "waste", it is not for everybody. WCAs have therefore introduced kerbside collection schemes for those householders who do not want to or are not able to compost their own garden "waste".

Centralised composting of garden "waste"

Many WCAs collect garden "waste" (often termed "green waste"), usually on a fortnightly basis. Some WCAs collect garden "waste" all the year round, others only during the spring and summer months. Most councils charge the householder for this service. Householders can also take garden "waste" to an HWRC for collection and many do.

Collected garden "waste" is taken to purpose-built, centralised composting facilities where it is treated as described in Chapter 5. The quality of the composted material depends entirely on the quality of the garden "waste" that goes into the process and its cleanliness. Any contamination of the inputs to the process will appear in the output.

Composted material from garden "waste" should not be confused with garden or potting compost sold through garden centres; it is of a much lower quality and there is a risk of contamination from non-garden "waste" materials. Consequently, it tends to be used for lower-grade, soil conditioning purposes, such as large-scale landscaping. These are still valuable end-uses for what was previously seen as "waste", but it is definitely a case of down-cycling. The lowest value use of this material is as an alternative to topsoil to provide daily cover to layers of landfilled "waste", as a landfill site is filled up.

Recovery of energy through Energy from Waste incineration

The residual Household Waste that is not collected separately for recycling or organic material recovery still has to be dealt with in some way. And let's face it, some consumers just won't bother to separate their unwanted stuff for reuse, repair or recycling, dumping it all into the "disposal" collection container. Also, some materials presented for recycling collection will be unsuitable for recycling and will be discarded at the MRF or reprocessing plant.

And there are some items such as broken crockery, disposable nappies, or filled dog waste bags which cannot be treated by recycling or organic material recovery.

Thus, the relatively small amount of Household Waste that would be left after recycling and organic recovery (see Chapter 11 for a discussion of the potential figures) should be collected and sent to an EfW incinerator to be burned and energy recovered from this process in the form of electricity.

We might think of EfW incinerators as a modern phenomenon, but they have in fact been around since Victorian times. Then they were called "destructors" which burned unwanted materials in large furnaces, sometimes even generating steam to power heating systems for swimming pools and to drive sewage pumping engines. A key driver for their popularity was a fear of the transmission of communicable diseases from infected clothes, mattresses and even wallpaper.[31]

Modern EfW incinerators have a more prosaic purpose, to reduce the volume of unwanted Household Waste and other "waste" and to recover energy from this incineration, as an alternative to landfill. The incinerators reduce the unwanted materials to hot gases and ash. The hot gases are captured and used to turn water into steam which is used to drive turbines to generate electricity. The exhaust gases then have to be "scrubbed" to remove toxic pollutants before being released into the atmosphere.

Even the ash from these incinerators can be used as low grade aggregate (after the metals it contains have been removed for recycling). It is only the by-products of the flue gas cleaning processes that have to be sent to landfill (see Chapter 5 for a description of the EfW process).

Whilst EfW plants turn "waste" materials into valuable electricity, this results in the loss of potentially valuable materials. If these materials are sustainably renewable, for example, food and garden "waste" or paper and cardboard (from sustainable managed forests), then this form of recovery is potentially sustainable, but this is not the case if the materials recovered are finite, such as in the case of plastics, which are derived from crude oil.

Early on in writing of this book, I was faced with a depressing possibility. If we can't find ways to increase recycling and organic recovery, the next best way to deal with the mountains of unwanted Household Waste is to burn it for energy generation in EfW incinerators. I was starting to think about every major conurbation having its own MRF and EfW plant. I found this prospect depressing, not least because:

- burning is no substitute for recycling (we destroy the materials and the financial and environmental value that has gone into producing and processing them; it is about as far from circular economics as you can get);
- EfW plants are not particularly energy efficient (15%–27%),[32] and
- there is an ongoing debate about gaseous emissions from such plants and what to do with the potentially toxic by-products they produce.

But by implementing the changes I have outlined above, namely: reducing consumption; increased reuse; reinventing repair; dramatically improving recycling; and much more organic recovery, we could significantly reduce the amount of materials for which there is no better solution than to burn them.

In 2017, there were 13.4 million tonnes of combustible materials discarded that were not collected for recycling or organic recovery.[33] If we take the capacity of a modern EfW incinerator as 200,000 tonnes per annum[34] then this would have required 67 EfW incinerators to dispose of it. So, we can perhaps see why so many EfW incinerators were being commissioned.

But, by implementing the changes I have listed above, the amount of combustible material that cannot be recycled or organically recovered drops dramatically to 5.5 million tonnes (see Chapter 11). Much of this material would be extremely difficult to recycle or recover as material, for example the Household Waste categories of treated wood (meaning painted or varnished), household hazardous "waste" such as paints and household cleaning products, and what are called miscellaneous combustibles (which stand out at a very significant 7.2% of total Household Waste).

Should we incinerate these materials rather than simply landfilling them? Is this a better way to treat them? Whilst it is true that we would get some energy recovery and size reduction, what about the pollution risks? If we do use EfW as part of the solution, implementing the reuse, repair, recycling and organic recovery changes I have outlined above would reduce the number of 200,000 tonnes per annum EfW plants needed from 67 to 28.

The problem of excess EfW incineration capacity

But in the words of the Apollo space missions: "Houston we have a problem", and it is potentially a big one, if it is not addressed. There were 48 EfW incinerators in operation in the UK in 2019, with six more being commissioned, 12 more in construction and three more planned for 2020, a total of 69 EfW incinerators and an expected increase in EfW capacity of 33%.[35] The existing EfW capacity in the UK in 2019 was 14.6 million tonnes per annum (tpa), projected to rise to 19.5 million tpa at the end of 2020. But under my proposals, we would only need a maximum of 5.5 million tpa of capacity or 28% of the projected capacity for Household Waste in 2020. What would all the additional EfW incineration capacity be used for?

Going right back to Table 1.1 in Chapter 1, there are other types of "waste" besides Household Waste that can and often are incinerated, for example commercial and industrial "waste" and construction "waste". Whilst the same attitude should be applied to these "waste" categories, that is, that they are not "waste" but unwanted materials, some will still be unsuitable for recycling or recovery and could be, and indeed already are, incinerated. Certainly, in the short term, if my proposed changes to the creation, collection and treatment of Household Waste were implemented, would there still be sufficient Household Waste to fulfil the contracts that WDAs have signed with the operators of EfW incinerators?

EfW incineration has effectively become the problem that landfill once was. It is cheaper than landfill, because of the Landfill Tax that now applies to landfilled Household Waste and because there are many EfW incinerators in existence – too many.

Now is the time to recognise the harm that EfW incineration is doing and will increasingly do to our recycling and organic recovery efforts (in the same way that landfill once did). I further think that the solution needs to be the same as it was for landfill.

The UK Government, or the devolved Governments, should impose an Incineration Tax, at least on Household Waste, to make EfW incineration a less financially attractive treatment option. In addition, I think the proceeds of such a tax should be used to partially fund the building of the recycling and recovery infrastructure that will be required to treat the much greater quantities of Household Waste that will need treatment, if my proposed product and packaging design changes and co-mingled collection approach were to be adopted. For example, the proceeds from such a tax could be used to stimulate the building of the MRFs that will be required and the additional material reprocessing capacity.

I cannot say strongly enough that the growth of EfW capacity in the UK is already threatening to, and will increasingly, prevent the advances in recycling and organic recovery that are possible. This is a key problem that we must address and the solution would appear to be straightforward.

Whilst in the very short-term we will need to use the EfW capacity that is already in existence or being built, we don't need any more. There needs to be an immediate moratorium on new EfW capacity and when existing incinerators reach the end of their lives, they should not be replaced. If we don't act on this (and I think this is one for central Government) then we risk recyclable and recoverable materials being sent for incineration simply because a WDA has signed a contract to supply a minimum tonnage of material every year to feed these monster facilities. Many people suspect that this is happening now in some parts of the country, whereby carefully separated recyclables are not being recycled, but burned. Let's hope this is just a conspiracy theory.

Given there are many WDAs that have minimum tonnage contracts in place to supply EfW incinerators with Household Waste, WDAs will have to work together to decide how best to fulfil these commitments, whilst allowing recycling and organic recovery rates to rise. Both commercial and construction "waste" might be used to fulfil any shortfall in Household Waste tonnages.

But setting aside the thorny issue of excess EfW capacity, let's not lose sight of the good news, which is that by implementing all of the changes I have advocated, no more than 25% of Household Waste would be destined for EfW treatment. The next chapter explains how this could happen.

Landfilling of waste

And now we've fallen off the bottom of the Materials Hierarchy. If we can't reuse, recycle or recover a material, we have no choice but to landfill it. Regrettable, but inevitable. Whilst we can strive to mimic nature by trying to use every output from one process as the input to another, sometimes what we have done to those outputs, particularly if they are complex products containing multiple materials, means they cannot be used for anything.

The amount of Household Waste being sent to landfill has already declined over recent years (Table 3.1 back in Chapter 3 shows 20% of Household

Waste was landfilled in 2015, falling to just 8% in 2019). This has been for three primary reasons:

- the amount of land available for landfill sites has declined, due to a smaller amount of suitable land being available, together with strong local opposition to the siting of landfill sites;
- the introduction of tighter environmental controls for the operation of landfill sites, in particular to prevent them from leaking toxic liquid effluent into the local groundwater and to manage the gas emissions from decaying material in a landfill site, has increased the costs of landfilling; and
- the introduction of the now punitive Landfill Tax making landfill an expensive option.

I'd like to think that a fourth reason is a general desire by WCAs and WDAs to recycle and recover rather than landfilling, but I actually think it is more a question of cost (see Chapter 7).

Again, the good news is that if we reuse, repair, recycle and organically recover what is possible and if we use existing EfW incineration plants to process the remaining 25% of Household Waste, then only 7% of Household Waste would have to go directly to landfill (see Chapter 11).

The nub of the problem

Many people see the solution to the problems we are facing as being to simply increase the rate of recycling (and they include organic recovery in this). As I will show in Chapter 11, this is not the case.

Others rightly argue that we should follow the Materials Hierarchy and accept that we can reduce, recycle and recover a significant proportion of our Household Waste, but that we will always need to dispose of some of it, either though EfW incineration or landfill or both. I want to challenge that assumption.

I simply don't accept that some things just can't be recycled or recovered. I agree with Tom Szaky that anything can be recycled if you put your mind (and money) to it, but I also disagree with him. His starting point is to take products and packaging as they are currently produced and to try to find a way to recycle them. I think we have to fundamentally change the way products and packaging are **designed** and hence manufactured, so that we bake reuse, repair, recycling and recovery into the life cycle of these products.

Instead of saying, "oh this product is too hard to recycle so we'll just burn it", let's redesign it so it **never** becomes "waste" and so that it can always be treated as an unwanted material(s) at the end of its useful life and be sent around the circular path to become useful once again.

But we also cannot say "let's recycle at any cost", this just isn't going to happen and nor should it. We will only start to address the consequences of our consumption if:

- producers design-with-the-end-in-mind;
- consumers buy-with-the-end-in-mind;
- WCAs enable householders to collect recyclable and recoverable materials easily and simply (KISS); and
- the waste management and reprocessing industries invest in the treatment facilities to process the collected materials.

But I want to go back to what I said early on in Chapter 1. Waste is a concept invented by humans; it simply doesn't exist in nature. The only reason we find ourselves with materials that we cannot easily deal with by the Three Rs is because we have made them too complex and economically unattractive to reprocess. We have sacrificed simplicity and sustainability for ease of manufacture, cheapness and convenience.

The key to unlocking the problem we face is **design**. We have to design products so they can be reused, repaired, recycled or recovered. By doing this we not only protect and preserve our precious planet, but we solve the growing problem of "waste" disposal; we simply won't need to dispose of "waste"; it's an outdated concept.

Having said this, sadly there **will** always be some "waste" that has to be disposed of as human activity does not mimic the natural world. But we could do a whole lot better, so let's start with this as we need to be pragmatic.

A word about waste water

Just about everything in this book is about how we need to treat discarded products and packaging, but there is one other category of things we throw away that should also be subject to the Materials Hierarchy and this is water.

We all know there are limits to the availability of clean tap water in this country, because whenever we have prolonged periods of warm, dry weather, we experience hosepipe bans to reduce the consumption of water, as the levels in our reservoirs and aquifers fall to unsustainable levels. And this will only get worse as we experience global warming.

For example, in the summer of 2022, drought was declared in eight of England's 14 regions. As a result, in some areas domestic taps ran dry.

But according to the National Infrastructure Commission in a report published four years earlier, this could soon happen to millions of us. According to the report there is about a one in four chance that at some point in the next 30 years we will face a drought so severe that a large number of households will have their water cut off[36] (and 2022 wasn't it).

This report made a number of recommendations including:

- improving the national water infrastructure, through a national transfer network in England and new infrastructure, such as reservoirs and water reuse systems;
- halving supply leakage by 2050 (in 2018, 20% of mains water was lost each day), which would save 1,400 million litres per day (according to *The Times* newspaper in 2022, 2,400 million litres of water are lost each day);[37] to put this in perspective, demand for water in 2018 in England and Wales was 14 billion litres per day...[38]
- reduce domestic demand from 141 litres per person per day to 118 litres per person per day (again according to *The Times* newspaper "British [water] demand – whilst one of the highest in Europe – has halved since the Seventies to 143 litres [per person] per day").[39]

This last point is particularly interesting, both because it alludes to what you and I can do to contribute to reducing the water shortage problem, but it also highlights the fact that our personal, day-to-day usage of water is only part of our environmental impact.

According to Rob Wilby, Professor of Hydroclimatic Modelling at Loughborough University, we need to use less of just about everything.[40] He wrote:

Water, like carbon, is embodied in all the goods and services that we consume ... Our typical daily water use of about 143 litres covers washing, cooking, cleaning, flushing the lavatory and outdoor purposes – but this is only what we see. According to the environmental charity WWF-UK, our average invisible water use is 4,645 litres per day [over 32 times our visible water usage].

For example, it takes up to 25 litres of water to produce a single sheet of A4 paper; a slice of bread requires 40 litres; and a pair of jeans 7,600 litres. To manufacture a car you need 150,000 litres.

This is an example of one of the hidden costs of consumption. Professor Wilby goes on to say "Much of this 'virtual water' is imported in goods – often from countries with less water than the UK. Through imports of staples such as meat, paper and plastics we are, in effect, supplementing our national water resources by billions of litres every day".

I'll leave the last words to Professor Wilby: "Just as we have become aware of the need to shrink our carbon footprint, we must also lighten our water footprint, regardless of when the drought breaks. Let's recognise the link between dry riverbeds and our everyday water use – both visible and invisible".

Water is thus another valuable resource that should be managed according to the principles of the Materials Hierarchy in terms of:

- reduce;
- recycle; and
- recover.

We can't really reuse or recycle tap water ourselves (but this is exactly what the water industry does), but we can reduce how much we consume by being more careful in how much water we use and by harvesting it, both tap water and rain water, for positive use rather than throwing it away (down the drain) and we can recover water by using it twice for different purposes.

Reduce water usage

You may think that it's really difficult to reduce the amount of water we use, but a few simple things can make a difference. Obvious examples are not leaving the tap running whilst you clean your teeth or running a tap until the water runs cold enough or hot enough; not emptying the kettle and refilling it every time you boil water for a cup of tea; or by collecting clean water that would otherwise go to waste (see below).

Some people argue that we can do even more, for example by washing our clothes less often than we do now, by only having four inches of water in a bath or by only flushing the toilet when it's really necessary (ever heard the adage "if it's yellow let it mellow, if it's brown flush it down"?) But to me these reduce the quality of our lives and certainly aren't things we need to consider yet (but global warming might change this). I'd rather stick to the simpler steps above, but I am really keen on recovering water as I explain below.

You can also reduce your mains water consumption by supplementing it with harvested rainwater. This water obviously hasn't been treated in the same way as mains water, so it's not as clean or hygienic, but it's great for watering the garden or washing the car, and it's free! I recently installed a water butt to collect and store rainwater and fitted a simple diverter from one of my rainwater downpipes. I have been amazed at how quickly the water butt fills up and how much water I can collect from just one downpipe. The water butt is also useful for storing water that I collect within the house (see below). It acts like a battery, storing "spare" water until I need it.

Recover water

As part of reducing our consumption of water we can recover water that would otherwise go to waste. Let me explain by starting with an example. When we want hot water from a tap or shower, we often have to let the water run for a period until the hot water comes through. Normally we let this clean, processed and valuable water just run down the plughole. But if we collect this water in a bowl or bucket, it can be used, for example for watering house or garden plants, for washing a car or for just about anything you can think of that needs water. And if you have a water butt, there is somewhere to store this water until you need it.

The opposite example is during hot weather, letting the cold tap run until the water is cold enough for a cooling drink. The better approach here is simply to fill a jug with cold water and put it in the fridge. You then have chilled water available to drink whenever you want it, without running off far more water than you actually need. What should you use to store the water in the fridge? Please don't go out and buy a plastic jug to do this (yet more plastic consumption). Now here's a radical idea, why not refill single-use water bottles and put them in the fridge?

These examples are about the recovery of clean water, but we can also recover used water, so-called grey water, from our washing-up or washing (either clothes or ourselves). This water has soap and detergents in it, but is still a valuable resource and is eminently suitable for watering the garden. So instead of tipping the washing-up water down the sink think about putting it on the garden (but, don't put it into a water butt where it can sit and fester over long periods).

But let's get back to the main subject of this book.

The materials we should consume

Currently, producers use whatever materials suit their purposes to manufacture products and packaging, without sufficient (if any) thought being given as to how these materials will be managed at the end of their useful lives. This has to change, so that producers only use materials that can be managed acceptably within the Materials Hierarchy.

Circular materials

As I explained in Chapter 6, very few materials are capable of being recycled in a circular manner. In fact, this only applies to metals and glass and, under the right circumstances, some plastics. Other materials, namely paper and cardboard, textiles and untreated wood can be recycled a limited number of times and in the case of textiles and untreated wood, this is down-cycling rather than recycling.

Renewable materials

As well as infinitely recyclable materials, there is also the category of renewable materials. I'm thinking here particularly of plant-based materials such as wood and natural fibre textiles. Making products and packaging from such materials does not fit the circular model (unless the materials are composted at the end of their lives and returned to the earth to grow more trees and plants). Plus, increasing consumption of such materials will require more land and water to grow them, both of which are in finite supply. So renewable materials are fine, provided there is enough land to grow them.

Material production methods

As well as the potential to recycle or recover materials, we also need to look at the production processes that are used to produce the materials in the first place, before they are turned into products.

Take cotton as an example. It is tempting to say we should use cotton (a renewable material), rather than man-made (plastic) textiles, but when we look at the environmental costs of producing cotton, this choice isn't as clear cut.

Did you know that it takes between 10,000 and 20,000 litres of water to grow and process one kilogram of cotton fabric (enough to make a pair of jeans and a shirt).[41]

Non-recyclable, non-renewable materials

The most obvious materials in this category are certain plastics and in particular plastic film. Whilst some plastics, notably Polyethylene Terephthalate (PET) and High Density Polyethylene (HDPE) bottles and rigid packaging are collected by WCAs for recycling, other polymers and all plastic film are not recycled, resulting in them being sent for EfW incineration or landfill.

The issue is one of being able to identify and sort different polymers so they can be reprocessed in single polymer streams, an impossible challenge for the householder and potentially the MRF operator.

How we should deal with "problem" materials

Let's start with packaging.

The problem is usually how the materials are used, rather than the materials per se. For example, when a number of materials are combined in composite packaging such as the layering of cardboard, plastic and aluminium in Tetra Pak style cartons, or cardboard and rigid plastic in blister packs or plastic film windows in cardboard cartons, then it is extremely difficult to separate these materials for recycling. Either the packaging is excluded by WCAs from recycling kerbside collection (as is usually the case with Tetra Pak style cartons) or one of the materials is treated as a contaminant in the reprocessing of the main material and is sent for disposal, as is the case of plastic windows in cardboard cartons.

The problem is more often than not the packaging of the product rather than the product itself. Whilst this is not always the case, it certainly is with food and other grocery products. This is an example of where three of the players in the supply chain all need to work together:

- producers need to replace the non-reuseable, non-recyclable materials used in their products and, in particular, their packaging, with materials that can be reused or recycled and/or they need to use a mixture of materials, but only if they are easy for the consumer to separate;
- retailers need to work with producers to achieve the above and promote those products that meet the highest environmental standards; and
- as consumers we need to do three things:

  - be alert to the materials used in packaging and ideally only buy those products that are packaged in what are reuseable or recyclable packaging;
  - separate combined material packaging into its component materials; and
  - if you feel you have no alternative but to buy what you think is a product made from, or whose packaging is made from, unsustainable materials, write to the producer and the retailer from whom you bought the product, explaining your concerns.

But I have a final word to say about changing the materials that are used to produce products and packaging. And this is that we must beware of the Law of Unexpected Consequences. By this I mean that if we make a change, it may affect other things that we haven't thought about and this might not be for the better.

For example, I could argue that all toothpaste tubes should be made from aluminium rather than plastic, as most are now (assuming that these aluminium tubes could be separated in a MRF and then reprocessed).

Now, in a truly circular economy, the aluminium would be recycled again and again. But as I've already explained in Chapter 6, no economy can ever be truly circular due to leakage, and if a market grows, there will be an increased demand for the toothpaste, meaning more new aluminium will be required. The mining of bauxite (aluminium ore) is extremely destructive to the environment where it occurs, so changing toothpaste packaging from plastic to aluminium would be attractive from a recycling perspective, but the unexpected consequence could be greater environmental destruction. Is bauxite mining any more environmentally destructive than drilling for and processing oil, from which plastic toothpaste tubes are made?

And in my earlier example, it takes between 10,000 and 20,000 litres of water to grow and process one kilogram of cotton fabric; that's a huge amount of water that has to be sourced and treated after use, if in fact it is (in many countries the water will just be dumped and fresh water sourced). But one company has developed an enzyme-based treatment for flax that produces a cloth that is equivalent to cotton, but only uses 17 litres of water per kilogram of material produced.[42] So is buying cotton products necessarily the best option from an environmental perspective?

We therefore need to be very careful when making changes to ensure that the overall environmental impact is actually positive.

We need an overall strategy and plan for change

I've talked about what we need to do differently within the context of the Materials Hierarchy and this will require change from all players in the Circular Supply Chain. To achieve such change, we need a clear strategy stating what we want to achieve and in broad terms how this will be done. We then need a clear plan of action which spells out what will be done, by whom and to what timescale.

This is where I think the final player in the Circular Supply Chain list comes in, namely the UK and devolved Governments. We need our Governments to implement national materials management strategies that incorporate all of the above changes. So-called Waste Management is a devolved area of responsibility within the UK so England, Scotland, Wales and Northern Ireland all act independently.

So, where are the UK Government and the devolved Governments today?

England

In 2018 Defra published "Our waste, our resources: a strategy for England".[43]

In February 2019 the Government published a consultation on measures to increase recycling from households and businesses to support the achievement of a 65% recycling target for municipal "waste" by 2035. The Government published a summary of its response to the consultation in July 2019.[44] This states that the Government will introduce measures for England to increase household recycling by requiring all local authorities to collect a consistent set of dry materials from households in England; to collect food "waste" separately from all households on a weekly basis; and to arrange for separate garden "waste" collection. These measures are expected to increase recycling from households from current levels to 65% by 2035. By 2035! Not a very ambitious target.

A key outcome of this consultation was Government proposals that all English WCAs collect a core set of Dry Recyclables: plastics, paper and cardboard, glass and metal cans on a fortnightly basis and food "waste" once a week. There is also a proposal to make garden "waste" collections free to householders with gardens.[45]

But what has also been proposed is that all English WCAs adopt a consistent approach to collection containers, with recyclables being separated by the householder into seven (yes seven!) bins. This has understandably been challenged by WCAs[46] as being highly inefficient and hence costly.

Whilst I fully support the proposals for the introduction of weekly food "waste" collection, I strongly disagree with the rest of this approach, both in terms of the lack of ambition for recycling and recovery rates and the timescales for their achievement (2035, honestly!) and the proposed collection methods (see Chapter 11 for what I think is possible).

In September 2023, the Prime Minister Rishi Sunak announced a number of changes to Government plans to row back on some of their environmental commitments; scrapping the proposals for seven bins was one of these,[47] but was it ever really going to happen?

There is nothing I can see in the strategy about:

- any ambition for Household Waste reduction, nor the encouragement for reuse or repair;
- the role of producers in designing for reuse, repair and recycling;
- how the treatment infrastructure will be created and funded to make these targets achievable; and
- the role of EfW incineration and how current capacity is to be reduced.

Scotland

The Scottish Government has set out an ambitious policy[48] as follows:
"We have several ambitious targets for reducing waste and increasing recycling. By 2025, we aim to:

- reduce total waste arising in Scotland by 15% against 2011 levels
- reduce food waste by 33% against 2013 levels
- recycle 70% of remaining waste
- send no more than 5% of remaining waste to landfill".

As you will see, these targets are very similar to what I propose in Chapter 11, except with regard to the target to reduce food "waste", which is laudable, but I think problematic. I assume that the term "recycle" refers to material recycling and organic recovery.

I also think the target of reducing "waste" arisings by 15% against 2011 levels to be rather ambitious if we take into account population increases, and there is nothing said about how this will be achieved.

I see nothing in the strategy about:

- the need to persuade producers to change how they design and manufacture products and packaging to make these targets achievable; the strategy appears to be aimed at how we treat "waste" differently, without addressing the barriers that exist to make this happen;
- how Household Waste is to be collected differently;
- how the treatment infrastructure will be created and funded to make these targets achievable; and
- the role of EfW incineration and how current capacity is to be reduced.

Wales

The Welsh Government has set the following even more ambitious targets: By 2025:

- 26% reduction in waste;
- zero waste to landfill;
- 50% reduction in avoidable food waste; and
- 70% recycling.

By 2030:

- 33% reduction in waste; and
- 60% reduction in avoidable food waste.

By 2050:

- 62% reduction in waste;
- zero waste (meaning no EfW incineration and no landfilling); and
- net zero carbon.

So, very ambitious targets. In 2021 the Household Waste recycling rate in Wales was 65.4% and Household Waste sent to landfill was 5%.[49] An impressive performance, but where did the other 30% go? EfW incineration?

The Welsh Assembly Government has mandated that Welsh WCAs use kerbside sorted collection, to maximise material quality, with householders being given quite a number of collection boxes to keep materials separate. This is a decision that I challenge on the grounds that many people will simply not separate materials to this degree; the imposition on the householder of having to store so many containers; and the inefficient use of collection vehicles as the separate compartments fill up at different rates (see Chapter 4 – Kerbside collection).

I can see no mention in the strategy of how these targets will be achieved, particularly in terms of:

- the need to persuade producers to change how they design and manufacture products and packaging to make these targets achievable; the strategy appears to be aimed at how we treat "waste" differently, without addressing the barriers that exist to make this happen;
- how the treatment infrastructure will be created and funded to make these targets achievable; and
- how current EfW incineration capacity is to be reduced (to zero).

Northern Ireland

Northern Ireland's last waste strategy "Delivering Resource Efficiency" was published in 2013 and a closure report was published in June 2022. A new draft strategy was expected in December 2022 and the final strategy is due to be published at the end of 2023. So, too early to comment.

What I would like to see in these strategies

To achieve the changes I think are necessary may require our Governments to legislate to make change happen. Governments don't like legislating, preferring to incentivise or allow industry to self-regulate. But because there are strong vested interests that could frustrate the required changes and because we need all the players in the Circular Supply Chain to act together to achieve common objectives, there will be some changes that will have to be mandated.

For example, I think there should be an outright ban on the exporting of UK Household Waste for reprocessing. We should take responsibility for cleaning up our own mess and do so to high environmental standards, which is not necessarily the case if we export our Household Waste to less developed countries.

Our Governments should also consider legislating to discourage the use of non-recyclable materials, such as expanded polystyrene.

And critically, I think we need an urgent moratorium on the building or expansion of further EfW incinerators, at least for Household Waste.

Whilst I said above that Governments don't like legislating, there are three pieces of legislation currently being prepared that have a direct relevance to what I'm proposing:

- proposed Bottle Deposit Schemes;
- Extended Producer Responsibility Obligations; and
- recycled plastic content targets in packaging.

Bottle Deposit Return Schemes (DRS)

England, Wales and Northern Ireland

There is a UK Government scheme proposed to cover these three regions, due to be launched in 2025. The scheme will cover only PET bottles and drinks cans.

Scotland

Scotland was due to introduce a mandatory Deposit Return Scheme from August 2023, but this has been delayed until 1 March 2024 to give all parties involved more time to prepare for the introduction of the scheme.[50] There is strong opposition to the scheme from Scottish businesses and some politicians and from the UK Government.[51]

The proposed scheme will include PET plastic bottles (not other plastic bottles such as PE), metal drinks cans and glass bottles. The deposit will be 20p per container.

Extended Producer Responsibility (EPR)

As I said earlier Extended Producer Responsibility requires producers to add all of the environmental costs associated with a product throughout the product life cycle, including its end-of-life treatment, to the market price of that product. But, in reality, these costs will not be borne by the producer, they will be passed onto the consumer, and quite rightly.

So what is the benefit of introducing EPR? I think it is that producers will be forced to recognise the costs associated, in particular, with the end-of-life treatment of their products and packaging and this might cause them to seek to minimise these additional costs that will be added to their products, in order for their products to remain competitively priced. This is another example of where cost drives behavioural change, rather than a concern for our environment, but it achieves the result that we need.

But the big benefit of introducing EPR is that it could create a source of financial funding to support the enhanced collection of packaging materials from households, through improved kerbside collection and the dismantling industry that I discussed earlier in this chapter (I discuss this further in Chapter 11).

Recycled plastic content targets in packaging

In June 2021, as part of the Finance Act 2021, the UK Government passed legislation to impose a tax on plastic packaging manufactured in or imported into the UK, that does not contain at least 30% recycled plastic. The aim of this legislation is to "... provide a clear economic incentive to businesses to use recycled plastic in the manufacture of plastic packaging ... this will stimulate increased levels of recycling and collection of plastic waste, diverting it from landfill or incineration".[52]

Certainly, we need increased levels of recycling and collection of plastic packaging "waste". In 2019, the collection rate of plastics packaging was 22%, if we ignore plastic "waste" exports for recycling, both consumer and non-consumer "waste".[53]

The tax has been set at £200 per tonne of chargeable plastic packaging components and came into force from 1 April 2022.[54]

Both pre-consumer and post-consumer plastic "waste" constitute recycled plastic.[55]

Will this make a difference? Well, any measure to increase demand for recycled materials is to be welcomed, but the cynic in me suggests that:

- by including pre-consumer "waste" plastic within the 30% recycled content, many producers will probably be able to comply now, without making any changes to their use of recycled plastic; it is the demand for post-consumer material that needs to be driven up, so this strikes me as a missed opportunity; and
- it is possible that some producers (or importers) will simply pay the tax and pass it onto their customers, thereby making no difference to the demand for recycled material.

Summary of what needs to change

In a nutshell, the problems we are facing are that:

- as consumers we are buying too much stuff;
- much of what we buy has **not** been designed by producers to be reused, repaired or recycled, but in future must be;
- the infrastructure for the kerbside collection of Household Waste needs to be changed to maximise the opportunities for reuse, recycling and recovery (particularly food "waste" management);
- we need more Dry Recyclables separation capacity and more reprocessing capacity to be built, located locally to where the unwanted materials are collected; and
- we need to create new industries to repair products and to dismantle products for reprocessing.

But there are clear solutions to these problems. I've summarised these above and have tried to present not only the problems we face, but what

I hope are practical, down-to-earth and, at times, blindingly obvious solutions to address the urgent and growing crisis facing our world; because we are quite literally trashing our planet.

How change could be achieved in practice

I appreciate there is a lot to take in in this chapter, so I have summarised below the key changes I think need to be implemented, based on the priorities of the Three Rs:

- **reduce** the amount of Household Waste generated by:
  - reducing our consumption through buying less and buying more carefully;
  - increasing direct packaging reuse; and
  - increasing product repair;
- increase the rate of **recycling** by:
  - convincing producers to "design-with-the-end-in-mind" and in particular to only use those materials that can be handled by the Household Waste management infrastructure, i.e. the kerbside collection, sorting and reprocessing arrangements;
  - making it simpler for householders to collect and set out packaging materials for recycling by providing them with a single co-mingled, Dry Recyclables wheeled bin;
  - encouraging the waste management industry to invest in appropriately located MRFs to take and sort the collected co-mingled materials (by appropriately located I mean local MRFs to serve local populations);
  - persuading the producers of packaging to:
    - reduce the range of materials they use to only those that are truly recyclable, which would require them to stop the use of composite materials such as drinks cartons and limit the plastics used in packaging to just two, PET and PE;
    - no longer use materials that are hard to separate into single materials, such as blister packs that combine glued cardboard and rigid plastic bubbles;
  - retailers taking back unwanted products and feeding those that cannot be repaired into a new dismantling industry, prior to the reprocessing of individual materials;
- increase the rate of **organic material recovery** by:
  - dramatically increasing the kerbside collection of food "waste";

- increasing the kerbside collection of garden "waste";
- encouraging the waste management industry to invest in appropriately located anaerobic digestion plants and commercial composting facilities to treat organic Household Waste; and
- expanding the existing limited product repair industry and creating a new dismantling industry to facilitate the recycling of products.

This list is a combination of attitudinal changes, such as how producers design products and packaging and how householders sort their Household Waste, plus significant infrastructure changes. The infrastructure changes comprise the building of many local MRFs, anaerobic digestion plants and composting facilities, plus product repair and dismantling facilities and creating more material reprocessing capacity.

But it also requires a major change to how much Household Waste is sent to EfW incinerators, which will potentially require legislation to place a moratorium on the building of more incineration capacity and to introduce a potential Incinerator Tax.

There is clearly a lot to be done. But what benefits will these changes bring?

Notes

1 Tom Szaky, *Outsmart Waste: The modern idea of garbage and how to think our way out of it*, Berrett-Koehler Publishers, 2014, p. 116.
2 When an unforeseen effect occurs, contrary to what was originally intended.
3 Sirin Kale, "I thought buying things would make me feel better. It didn't: the rise of emotional spending", *The Guardian* Online, 9 February 2021, accessed 17 March 2023.
4 Glenn Adamson, *Fewer, Better Things: The Hidden Wisdom of Objects*, Bloomsbury Publishing, 2018.
5 Glenn Adamson, *Fewer, Better Things*, p. 137.
6 Rosie Kinchen, "Greta Thunberg on turning 18 and why she won't tell you off for flying", the *Sunday Times*, 2 January 2021.
7 "Organic jeans for rent: MUD Jeans", featured on the ellenmacarthurfoundation.org website, accessed 4 October 2021.
8 Peter Lacy & Jakob Rutqvist, *Waste to Wealth*, Palgrave Macmillan, 2015, p. xvii.
9 "Cars parked 23 hours a day", www.racfoundation.org, 8 July 2021, accessed 21 March 2023.
10 Lauren Bravo, *How to Break Up with Fast Fashion: a guilt-free guide to changing the way you shop – for good*, Headline Home, 2020.
11 Lauren Bravo, *How to Break Up with Fast Fashion*, pp. 4–5.
12 Lauren Bravo, *How to Break Up with Fast Fashion*, pp. 2–3.
13 Parliamentary business/Publications & records/4 Textile Waste and Collection, www.parliament.uk, accessed 23 March 2023.
14 Lauren Bravo, *How to Break Up with Fast Fashion: a guilt-free guide to changing the way you shop – for good*, Headline Home, 2020, p. 15.
15 "Scotland's deposit return scheme", www.gov.scot/news/scotlands-deposit-return-scheme, 14 December 2021, accessed 22 March 2023.

16 John Boothman, "Drinks firms want to wait for UK recycling scheme", *The Times* newspaper, 28 May 2023.

17 Defra Press Office, "Coverage of plans to introduce a Deposit Return Scheme", 23 January 2023, accessed 22 March 2023.

18 This was the target set for the UK for 2020 by the European Commission when the UK was part of the European Union.

19 Tom Whipple, *The Times* newspaper, "The latest green plea? Put less in the recycling", 31 May 2023.

20 Welsh Assembly Government, "Towards Zero Waste – The Overarching Waste Strategy Document for Wales", June 2010, p. 20.

21 WRAP, prepared by Eunomia Research Consulting Ltd, "National Household Waste Composition 2017", Table 7, 2019.

22 "Household collections with FlexCollect", www.flexibleplasticfund.org.uk, accessed 27 February 2023.

23 British Metal Recycling Organisation (BMRA), "WEEE doesn't go in the bin!", recyclemetals.org, accessed 4 September 2022.

24 "Lithium-ion Battery Waste Fires Costing the UK over £100m a Year", www.euno mia.co.uk,13 January 2021, accessed 6 October 2022.

25 WRAP, "Food Surplus and Waste in the UK – Key Facts", January 2020, p. 1.

26 WRAP, "Food Surplus and Waste", pp. 2 and 13.

27 Colin Butfield and Jonnie Hughes, *Earthshot – How to Save Our Planet*, John Murray Publishers, 2021, pp. 319–320.

28 This is 6.6 million tonnes out of the total Household Waste figure of 27.3 million tonnes (see Chapter 2).

29 WRAP, "Food Surplus and Waste in the UK – Key Facts", January 2020, pp. 1–2.

30 WRAP, "Food Labeling guidance", Nov 2019, pp. 19–22.

31 Emily Cochrane, *Rummage – A History of The Things We Have Reused, Recycled and Refused to Let Go*, Profile Books Ltd, 2021, pp. 115–116.

32 Energy Saving Trust, "Generating energy from waste: how it works", blog post, 24 January 2019, accessed 6 March 2023.

33 WRAP, prepared by Eunomia Research Consulting Ltd, "National Household Waste Composition 2017", 2020, author analysis of Table 7.

34 The recently opened Javelin Park EfW plant near Gloucester has a capacity of 190,000 tpa.

35 Tolvik Consulting, "UK Energy from Waste Statistics – 2019", May 2020, author's analysis of Figures 37–40, pp. 18–19.

36 National Infrastructure Commission, "Preparing for a drier future – England's water infrastructure needs", April 2018, p. 3.

37 *The Times* newspaper, Environmental Notes email, 16 August 2022.

38 Defra, "Water supply and demand management", , National Audit Office, June 2020, p. 4.

39 *The Times* newspaper, Environmental Notes email, 16 August 2022,

40 Professor Rob Wilby, "How Britain can keep the taps running", the *Sunday Times*, 13 August 2022.

41 House of Commons Environmental Audit Committee, "Fixing Fashion: clothing consumption and sustainability", Sixteenth Report of Session 2017–19, February 2019, p. 29.

42 Peter Lacy & Jakob Rutqvist, *Waste to Wealth*, Palgrave Macmillan, 2015, p. 38.

43 Defra, "Our waste, our resources: a strategy for England", 2018.

44 Defra, "Consistency in recycling collections in England: executive summary and government response", 2019.

45 Defra, "New plans unveiled to boost recycling", press release, 7 May 2021.

46 "Bin collections: Plans to change recycling risks chaos, say councils", BBC News website, 20 March 2023.
47 Anthony Reuben, "Seven bins and Sunak's other net zero claims fact checked",www.bbc.co.uk/news/uk-66878893, accessed 28 September 2023.
48 "Managing waste" www.gov.scot/policies/managing-waste/, accessed 19 April 2023.
49 "New stats show Wales upholds world class recycling rates, despite pandemic", www.gov.wales, press release, 18 November 2021, accessed 19 April 2023.
50 Calum Watson, "Why has Scotland's deposit return scheme been delayed?", BBC Scotland News website, 18 April 2023, accessed 19 April 2023.
51 "Scotland must rethink bottle recycling scheme – UK minister", BBC News website, 12 February 2023, accessed 18 April 2023.
52 "Introduction of Plastic Packaging Tax from April 2022", www.gov.uk, 20 July 2021.
53 WRAP, "Plastics Market Situation Report", 2021, p. 11.
54 "Rate", Clause 45 Finance Act 2021.
55 "Meaning of 'plastic' and 'recycled plastic'", Clause 49 Finance Act 2021.

11

WHAT WE COULD ACHIEVE IF WE CHANGED

Finally, let's turn to what we could potentially achieve if we implemented the changes I am advocating and the realistic targets we should set for our future management of unwanted materials.

First, let me take you right back to where we started. The most recent Government published data show that we currently generate about 27.5 million tonnes of Household Waste in the UK every year.

And again, I'd just like to remind you of the Three Rs: reduce (including reuse and repair), recycle and recover (both organic materials and energy).

The latest reported UK combined recycling and organic recovery rate (2020) for Household Waste was 44.4%.[1] This was very slightly up from 44.3% in 2018[2] and from 40.4%[3] in 2010, but down from 46.0% in 2019.

This represented a 12% increase in the reported UK Recycling Rate over ten years. Not a very impressive increase in performance. And remember, the Government's stated "Recycling Rate" is actually a combination of material recycling (26%) and organic material recovery (18%).

We need change – it needs to be radical and it needs to be rapid.

The key changes required

In the previous chapter, I set out my proposals for the way Household Waste is collected and treated:

- product changes:
  - design and manufacture for: reuse, repair and recycling (design-with-the-end-in-mind);

DOI: 10.4324/9781003504757-12

- implement revised collection arrangements for unwanted products; and
- expand the existing product repair industry and create a new product dismantling industry to separate materials for reprocessing;

- packaging changes:

  - design for reuse or recycling;
  - radically reduce the number of plastics used in packaging to simplify recycling and specifically, limit plastic film to HDPE/LDPE (polythene) film only;
  - all Dry Recyclables collection to be by co-mingled collection of a UK-wide standard list of materials in a single wheeled bin per household;
  - these co-mingled materials to be separated in a local MRF; and
  - this standardised list of materials to include plastic film;

- organic "waste" changes:

  - food "waste" to be collected from ALL households and treated in local anaerobic digestors; and
  - all unwanted garden "waste" to be collected and locally composted (although some householders will prefer to compost garden "waste" themselves).

The data

The baseline data, labelled "2017" in the following charts are taken from the most authoritative analysis of Household Waste that was carried out by WRAP six years ago.[4] This analysis reported by individual material category, for example by paper, cardboard, glass, plastics, how much of each material was separated for recycling/recovery at the kerbside, at HWRCs and through bring collection, as well as how much of these materials were **not** separated for recycling/recovery and so were sent for disposal.

In order to analyse the impact of my proposed changes on the recycling, organic recovery and energy recovery rates, I have assumed that all potentially recyclable packaging materials would in future only be collected for recycling via kerbside collection, directly from households. I have also assumed that the following **products** would be collected and dismantled for material reprocessing:

- all WEEE (both large and small);
- clothing, textiles and shoes; and
- household batteries.

I have assumed that all recyclable packaging materials, i.e. paper, cardboard, glass, etc. that were **not** separated for recycling in 2017 and so were sent for disposal, would in future be separated for recycling through kerbside co-mingled collection of Dry Recyclables, plus organic materials would be collected by the kerbside collection of food "waste" and the kerbside collection of garden "waste".[5] In other words, all of the recyclable/ recoverable materials would be collected in one of three containers at the kerbside. The Dry Recyclables would then be sorted and separated in a local MRF, the food "waste" would be processed in a local anaerobic digestor and the garden "waste" would be composted in a local composting facility.

I have also assumed that the Dry Recyclables I have listed would only be collected via kerbside collection and that bring collection of these materials at HWRCs and at stand-alone bring sites would cease, as kerbside collection would make these unnecessary.

So, let's look at the potential effect of making each of these changes.

The impact of these changes

I'd like to make one thing clear before I discuss the potential in detail. All the numbers I will quote represent the tonnages of materials collected for recycling or organic recovery. This is not the same as the amount of materials that will actually be recycled/recovered, because of losses in the recycling/ recovery processes. So, when I talk about a recycling rate or recovery rate, this refers to the **collection** rate, not the amount of material actually recycled/ recovered. I address this further under the section "A dose of realism".

If we start at the top of the Materials Hierarchy and work down, we can't actually measure the impact of reduction (but I'll come back to this when I talk about targets). So, the following analysis shows the impact these changes would have on the amounts of Household Waste treated by organic recovery and recycling. But please note, this is the best case scenario (I look at what reality might deliver in a minute).

The impact of making the proposed changes

Figure 11.1 shows three scenarios across the bottom:

- the 2017 baseline numbers,[6]
- the impact of implementing the organic material changes only; and
- adding the impact of making the Dry Recyclables changes.

I've shown the effect of the organic recovery changes first, as these are the simpler changes to make and every WCA was required to introduce them during 2023 (but many didn't). I've then added the impact of the proposed recycling changes, which will take a little longer to implement.

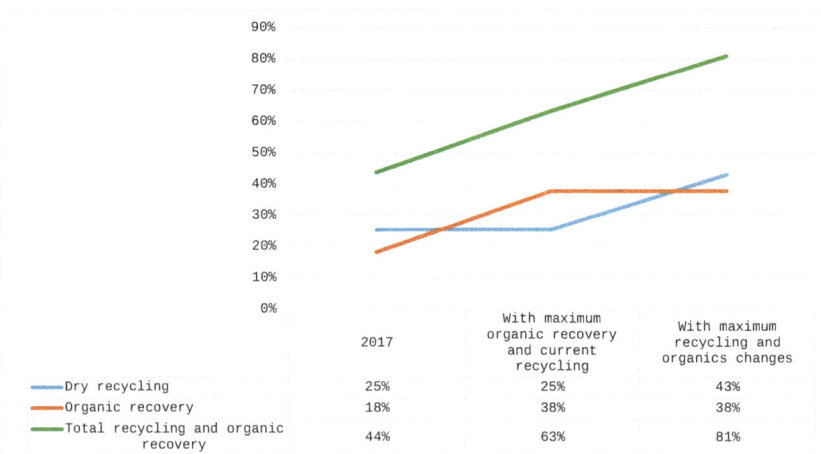

	2017	With maximum organic recovery and current recycling	With maximum recycling and organics changes
Dry recycling	25%	25%	43%
Organic recovery	18%	38%	38%
Total recycling and organic recovery	44%	63%	81%

FIGURE 11.1   The impact of making the organic recovery and recycling changes.

The orange line shows the recovery rate for organic materials. The impact of making the changes I've proposed, in particular food "waste" collection and processing, is significant. The organics recovery rate would increase from 18% in 2017 to 38%, more than doubling.

The blue line shows that making my proposed changes to Dry Recyclables and product collection and treatment would increase the recycling rate from 25% in 2017 to 43%, a 72% increase.

The green line shows the combined improvements in the total recycling and recovery rates. Just making the organics recovery changes would increase this rate from 44% in 2017 to 63% and adding the recycling changes would boost this combined rate to 81%, almost doubling the 2017 rate.

These are impressive and exciting numbers. But I don't want us to get carried away and will inject some reality in a moment. But before I do, I'd like to look at the impact these changes would have on EfW recovery, as shown in Figure 11.2.

The impact on EfW recovery

I have assumed that all combustible material that was not recycled or recovered as organic material in 2017 was sent for EfW recovery, i.e. anything that could burn, was burned. This is of course a gross simplification and some of this material would have been sent for landfill.

Clearly, increasing organic recovery and Dry Recyclables and product recycling will reduce the amount of Household Waste being sent for energy recovery through EfW incineration. Figure 11.2 quantifies this impact.

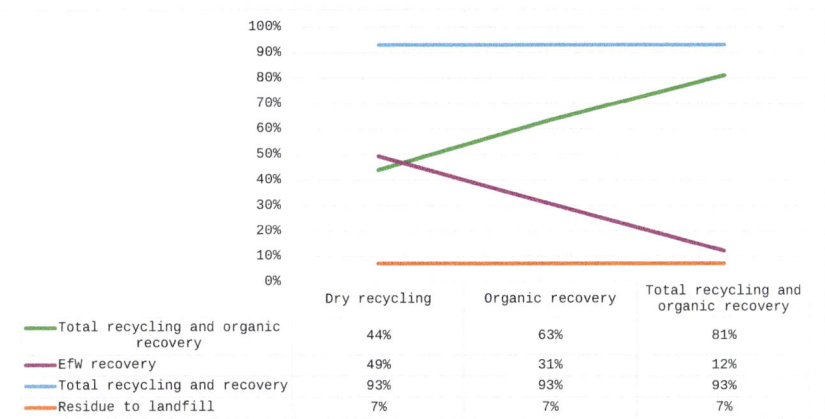

	Dry recycling	Organic recovery	Total recycling and organic recovery
Total recycling and organic recovery	44%	63%	81%
EfW recovery	49%	31%	12%
Total recycling and recovery	93%	93%	93%
Residue to landfill	7%	7%	7%

FIGURE 11.2   The impact of making the proposed changes on EfW recovery.

The green line, i.e. the total organics recovery and recycling rate, is the same as before, but I have added a purple line which shows the effect the organic recovery and recycling changes would have on the amount of Household Waste being sent for EfW incineration.

Maximising organic recovery would reduce the overall tonnage of Household Waste being sent to EfW incineration from 49% of all Household Waste to 31%. If we then add the recycling changes, the EfW figure falls to just 12% or 3.3 million tonnes. This would have a dramatic effect on EfW tonnages arising from Household Waste which I discussed in Chapter 10 under the heading "The problem of excess EfW incineration capacity".

The blue line in Figure 11.2 shows the combined recycling, organic recovery and energy recovery rate, i.e. all forms of recycling and recovery. The organic recovery changes would have no impact on this combined figure as the previously incinerated organic materials would now be recovered by anaerobic digestion and composting. There would also be no change in this rate from introducing the recycling changes, as I have assumed any non-recycled, combustible materials would have been sent for EfW recovery, rather than landfill.[7]

The upshot is that the overall recovery and recycling rate would not change with just 7% of Household Waste going directly to landfill, but far less recovery would be by EfW incineration. And remember, EfW incineration is only slightly better than landfill as a treatment method, so the important figure to focus on is the combined recycling and **organic** recovery rate, which as I've said would nearly double from 45% in 2017 to 81% of total Household Waste treated.

A dose of realism

But, all of the above is based on us achieving 100% of the changes I have proposed. Clearly, this isn't realistic, as it will take time to make these changes and, if we're honest, not every householder will correctly sort and separate their unwanted materials, even though the arrangements will be less complicated than now.

In addition, I said at the start of this chapter that the recycling and recovery rates I am quoting are for collection only and do not allow for the rejects from the recycling and recovery processes. Clearly, these will reduce the actual recycled and recovered tonnages achieved.

So, what might we achieve if we were less than 100% successful?

Figure 11.3 shows three scenarios: a 100% success rate; a 95% success rate; and a 90% success rate.

The combined organics recovery and recycling rate, if all the changes were made, is shown in green and the impact on the EfW recovery rate is shown in purple. The overall recovery and recycling rate (including EfW energy recovery) is shown in blue and the amount of Household Waste being sent directly to landfill is shown in orange.

I've already discussed the 100% success rate figures above, resulting in 81% recycling and organics recovery, 12% EfW recovery and 7% directly to landfill. If we were to only achieve 95% of the proposed changes, these numbers change to 77%, 16% and 7% respectively. And if we were to only achieve 90% success, the numbers would be: 73% recycling and organics recovery, 20% EfW recovery and 7% directly to landfill.

Two things are clear from Figure 11.3:

- the less successful we are in implementing my proposed changes, the lower the combined organic recovery and recycling rate (the green line); but

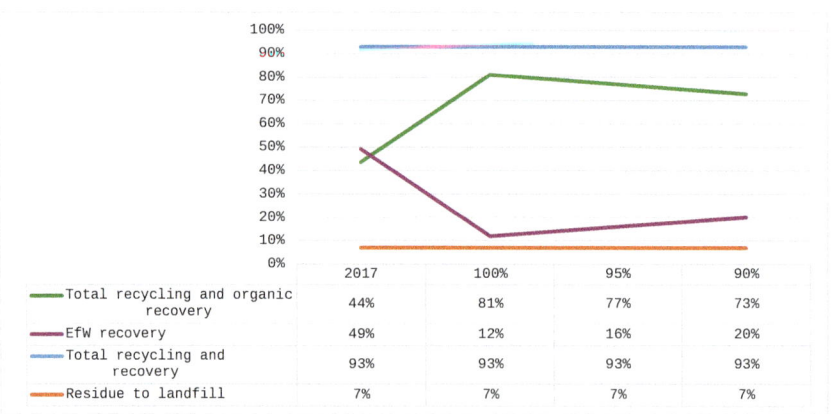

	2017	100%	95%	90%
Total recycling and organic recovery	44%	81%	77%	73%
EfW recovery	49%	12%	16%	20%
Total recycling and recovery	93%	93%	93%	93%
Residue to landfill	7%	7%	7%	7%

FIGURE 11.3   Success scenarios and a dose of realism.

- because the materials not recycled or recovered as organic materials are suitable for incineration, the overall recovery and recycling rate would not reduce.

Being realistic, I think it is this last set of numbers (the 90% success rate) we should focus on as representing our short-term ambitions, but the 95% success rate should be our longer-term goal.

What we could realistically achieve

This means that implementing these changes to the 90% success level would boost recycling to 39% (up from 25% in 2017), organics recovery to 34% (up from 18%) and EfW energy recovery down from 49% to 20%. This would move the current combined organics recovery and recycling rate from 44% in 2017 to 73%, so a very marked and necessary improvement. And I think we can take the 90% success level as representing the tonnages of materials that would actually be recycled/recovered, not just collected.

The risk of spurious accuracy

There is a concept in science and engineering called "spurious accuracy" when data or numbers are presented as being more accurate than they should be. Given that the above figures are based on a theoretical analysis of a limited study of Household Waste carried out six years ago, and an assumed success rate of 90%, from now on I will adopt more rounded numbers as follows:

- recycling: 40%
- organic recovery: 30%
- combined recycling and organic recovery: 70%
- EfW recovery: 25%
- landfill: 5%

Could we really achieve a recycling and organic recovery rate of 70% compared with 44% in 2017 and 45% in 2021? Yes, I think we could, because if we break down the changes required they are all eminently doable.

First, the simplest change with the biggest impact would be kerbside food "waste" collection (which has already been mandated by Government);

Second, the big changes in terms of recycling would be:

- producers to design for reuse and recycling including changing some of the materials they currently use, primarily in packaging, to only use truly reuseable or recyclable materials, but this is not rocket science;

- retailers and producers to facilitate the collection of unwanted products either for repair or dismantling and to create an expanded repair industry and a new dismantling industry;
- WCAs and WDAs to collect and contract the processing of kerbside collected packaging materials in a different way, i.e. the collection and sorting of co-mingled materials in a MRF; this is a big change, but happens now in some UK local authorities and happens in many countries in the Western World, particularly in Europe and the USA;
- consumers to willingly separate their recyclable materials, but into much simpler containers (KISS) and to take their unwanted products to retailers for subsequent treatment;
- the waste management industry to invest in new, local MRFs to accept and separate the co-mingled materials; and
- the material reprocessing industries to gear up to receive much greater quantities of collected and sorted materials for reprocessing by investing in new, strategically located reprocessing plants.

How the required new infrastructure would be funded is discussed below.

In summary

Just to bring the above analysis together, if:

- we implemented the organic recovery and recycling changes I am proposing; and
- we achieved just 90% of what is theoretically possible; then

  - the recycling rate would increase to 40% (up from 25%);
  - the organics recovery rate would increase to 30% (up from 18%); and
  - the EfW energy recovery rate would fall from 49% to 25%.

This would move the current combined recycling and organics recovery rate from 44% to 70%, and the total recycling and recovery rate to 95% with only 5% of Household Waste going directly to landfill. I hope you agree that these are pretty exciting possibilities, but let's not get carried away. I stress these are theoretical figures, assuming all the stars are aligned.

The infrastructure required

If we are to achieve these improvements, we need three major attitudinal and operational changes to happen:

- limiting the range of materials used by producers, particularly in the case of consumer packaging and designing-with-the-end-in-mind;

- changes in the way Household Waste is collected, principally implementing a simple method of collection of a standard range of Dry Recyclables in single, co-mingled wheeled bins and the collection of food "waste" from **every** household; and
- changes in the attitudes of consumers and householders to how they see and deal with their unwanted materials, in essence being willing to separate Dry Recyclables and food and garden "waste" from other unwanted materials for kerbside collection and taking back unwanted products to retailers for repair or dismantling.

But in addition to these new ways of working and thinking, we will need to expand the UK infrastructure for actually dealing with our unwanted materials.

This goes beyond recycling and organic recovery, to include the two critical elements of reduction, namely, reuse and repair. So, we need to expand our existing, limited treatment capacity and put in place a nationwide infrastructure to support reuse, repair, recycling and organic material and energy recovery.

So, taking each of these in turn.

Reuse

The infrastructure required to support direct packaging reuse is very, very slowly starting to appear. Obviously dealing with reuseable glass milk bottles has been going on for decades and is easy to expand to accommodate the increasing number of consumers who are switching from plastic to glass milk containers. I say easy because the dairy industry knows how to do it, has been doing it for a long time and it makes economic as well as environmental sense.

And whilst it is good that retailers, supported by a small number of producers are starting to trial reuse and refill schemes, what is being done is too little and far too slow. We need all retailers, both large and small, to introduce a range of reuse schemes, supported by all producers, not just some, potentially using standardised packaging containers, such as reuseable glass bottles. There is too much greenwashing and not enough action.

Repair

We don't have a serious national capability to repair products. Some shops do exist that will undertake repairs, such as: small, local tailoring businesses for clothes and cobblers for shoes; small, local shops, but also some larger chains, who will repair clocks, watches and jewellery; and far fewer shops that will undertake the repair of electrical and electronic products.

I see this latter product group as a major economic growth opportunity, creating new, highly skilled jobs and preserving useable products, rather than discarding them, thus saving materials and reducing "waste". BUT, and this is the big but, whilst this can happen to some extent now, it will require producers to change the way they design and manufacture electrical and electronic products to facilitate repair. Plus, they will need to provide spare parts to this new repair industry. Again, it comes back to producers designing products to make them suitable for repair.

Recycling and organic recovery

In terms of packaging recycling and organic material recovery, every WCA will need to be able to deliver collected:

- Dry Recyclables to a local MRF;
- food "waste" to a local anaerobic digestion plant;
- garden "waste" to a local composting plant;
- Household Waste for disposal to a local EfW incinerator; and
- materials that are unsuitable for any other form of treatment, including EfW incineration, together with the residues from recycling, organic recovery and EfW incineration, to a local landfill site.

This means that we need many more of these facilities, spread throughout the UK and conveniently situated for WCAs to be able to deliver to them economically.[8] In addition and critically, we need more material reprocessing plants, located within an economical transport distance of the MRFs that will generate the separated recyclable materials (which is key to stopping the export of recyclable materials).

Which raises two questions:

- are there enough MRFs, anaerobic digestors, composting plants and EfW incinerators to deal with the volume of Household Waste being generated, if we were to adopt the methods of treatment I am proposing? and
- are these facilities in the right places, to make it economic to transport the materials to them?

The simple answer to both of these questions is "no" (apart from EfW incinerators), but let's dig deeper.

The answer to the first question is that there are clearly not enough MRFs (of the right type), anaerobic digestion plants and composting sites in existence in the UK. These will need to be built and they will need to be sited close to the source of the Household Waste they are treating, to minimise transport impacts. As I said in Chapter 5, such facilities are relatively small in scale and so are suited to being constructed at a local level. As Table 11.1 shows, we will need of the order of:

- 200+ MRFs;
- 80+ anaerobic digestion plants; and
- 190+ composting sites.

Equally, as explained in Chapter 1, there is insufficient capacity to reprocess some recyclable materials, which is why we have the ridiculous situation of some WCAs sending recyclables overseas for apparent reprocessing.

What we won't need is more EfW incinerators; we have too many of these plants now and will need significantly fewer in the future as reduction, recycling and organic recovery become the norm (see Chapter 10 for the discussion on excess incineration capacity).

As I said way back in Chapter 1, I think the UK (and indeed every country) should be self-sufficient in the treatment of its own unwanted materials (in essence taking responsibility for its own mess), which will involve the construction of local MRFs; anaerobic digestion plants; and composting facilities.

The same argument about scale is not true of the Dry Recyclables **reprocessing** plants, such as paper mills and glass making plants, which operate at a much larger scale. Obviously, some of these exist already and will continue to take in separated single stream materials, but there is insufficient reprocessing capacity in the UK for some materials, such as paper, which is why some Dry Recyclables are exported for reprocessing. The additional facilities needed should be built in the right geographical locations to match the levels of materials arisings.

Table 11.1 summarises the infrastructure that will be needed to reprocess collected materials (based on the 90% success rate). This will be the most capital intensive and one of the four critical changes required in our treatment of Household Waste (the other three being: a standardised collection of co-mingled Dry Recyclables; the separate collection and treatment of food and garden "waste"; and the collection and treatment of household products).

Obtaining accurate data on how many processing facilities already exist, their locations and the materials they accept is very difficult. For example, many plants, such as paper mills and anaerobic digestion plants take in materials from a number of sources with Household Waste derived materials being only one; identifying how much of their input is from collected Household Waste and assessing how much capacity they have to take more of these materials is beyond the scope of this book.

In Table 11.1, I have shown how much capacity would be required to reprocess **all** of the materials that would be collected if we achieved 90% of

TABLE 11.1 The infrastructure required for success

	Arisings (million tpa)	Typical plant capacity (tpa)	No. facilities required
Organics to be treated:	**9.3**		
• anaerobic digestion	4.5	55,000[a]	82
• composting	4.8	25,000	191
Dry recyclables to be MRF sorted	**10.6**	50,000[b]	212
Dry recyclables to be reprocessed:			
• Paper	3.7	100,000[c]	37[d]
• Cardboard			
• Glass	1.6	Unknown	Unknown
• Aluminium	0.7	N/A	Existing
• Steel	0.3	Unknown	Unknown
• Rigid plastic	1.1	20,000[e]	95
• Plastic film	0.8		
• Textiles and shoes	0.3	Unknown	Unknown
• WEEE	0.1	Unknown	Unknown
• Household batteries	0.6	Unknown	Unknown
• Non-treated wood	0.4	Unknown	Unknown
Total Household Waste to be treated	**19.9**		
Household Waste to be incinerated	**5.5**	190,000	29
Household Waste to be landfilled	**1.9**	N/A	N/A
Total Household Waste	**27.3**		

Notes:
[a] Estimate.
[b] This is a large-scale MRF.
[c] Soren Back, "The British Paper Industry Today", paperadvance.com, 30 March 2021, accessed 2 August 2022.
[d] This is the number of mills required just to reprocess collected, unwanted paper and cardboard, if that was all they did.
[e] An indicative figure only.

the potential collection amounts. Where I have been able to identify data on current capacity I have identified this in the notes to the table.

Funding the infrastructure required

The role of Extended Producer Responsibility

As I have said above, not only do we need a very significant number of local sorting plants (MRFs), organic "waste" treatment plants (anaerobic digestion and composting), but we also need many more regional material reprocessing plants, plus expanded repair facilities and a new product dismantling industry. So how are these to be funded?

Paying for the treatment of kerbside collected materials

The normal way the costs of Household Waste treatment plants are recovered is through gate fees charged to the WCA/WDA, for example for EfW incinerator and landfill.

The same would apply to local MRFs and organic material treatment plants that will/do charge a gate fee for every tonne of Household Waste that they accept. This is paid for by the WDA (which funds this through the Community Charge), in the same way it pays for EfW incineration or landfill. Provided the plant gate fees are comparable with or less than EfW and landfill gate fees, there would be no increase in the costs to the WDA.

But where MRFs and, to a lesser extent, anaerobic and composting plants differ from other treatment plants is that they sell the materials they produce, thus reducing their operating costs. But the market prices paid for the outputs from these plants may be too low to make the operating costs add up, particularly for MRFs.

I say particularly for MRFs, as there are further costs associated with the reprocessing of the separated Dry Recyclables, when the sorted materials are reprocessed and again sold. If the prices paid for the reprocessed materials are not high enough, the operating costs of a MRF and the material reprocessing costs may be higher than the combined income from the sale of the reprocessed materials. This is one reason why so much collected Dry Recyclables are exported to countries with lower operating costs. MRFs will therefore have to charge a gate fee to accept kerbside collected Household Waste.

This is where we come back to the concept of Extended Producer Responsibility (EPR). As I explained in Chapter 10, under "EPR", producers of UK packaging and retailers of imported packaged products will have to add the full costs of treating the "waste" generated by their products, starting with packaging in 2025. This will create funding to enhance the collection and treatment of packaging and products, which I argue should be used to fund:

- the MRF sorting of co-mingled collected Household Waste;
- the dismantling of collected unwanted products; and
- subsidising the reprocessing costs of the collected and separated materials.

However, we can't wait until 2025, so I suggest that the higher costs of collecting and treating Household Waste through recycling and organic recovery could be further subsidised by:

- repurposing the proceeds from the Landfill Tax; and/or
- introducing and hypothecating a Household Waste Incineration Tax.

Repurpose some or all of the proceeds from the Landfill Tax

Currently, the proceeds of the Landfill Tax are used to fund environmental improvement projects that offset some of the negative impacts of living in the vicinity of a landfill site for affected communities. This is done through not-for-profit organisations that deliver projects for the benefit of communities and the environment in the vicinity of a landfill site.

As an alternative, consideration could be given to using some of these collected monies to help fund the building of the local infrastructure required comprising MRFs, anaerobic digestion plants and composting facilities.

Introduce and hypothecate a Household Waste Incineration Tax

As I have said in Chapter 10, I think the UK Government should impose an Incineration Tax on Household Waste, to correct the problem of EfW incineration being a cheaper treatment option than some kinds of recycling and organic recovery.

If this were done, then some or all of the proceeds of the tax could be hypothecated (ring-fenced) to partially fund the creation of the required recycling and organic recovery infrastructure.

Conclusions on funding

Clearly, introducing the methods of collection and treatment of Household Waste by recycling and organic recovery will increase the costs of collection and treatment.

I have listed two ways that these cost increases should be funded:

* through the proceeds raised from the introduction of Extended Producer Responsibility on packaging (and ultimately on products); and
* from changing the way some of the proceeds of the Landfill Tax are used and the introduction of an Incineration Tax.

Dare I suggest that funding the new collection and treatment methods I have proposed will therefore not be a problem for local government?

Given the proposals I have made for the revised collection and treatment of Household Waste and that these would appear to be fundable, what targets should we set ourselves?

What our targets should be

History

The UK had a combined recycling and organic recovery target of 45% by 2020. But as I've already said, this target was made up of two very different elements: Dry Recyclables recycling and organics recovery. The target was achieved in 2020 with a Dry Recyclables recycling rate of 26% and an organics recovery rate of 18%.

But such targets and the measurement of their achievement are flawed for two reasons:

- two very different methods of treatment, with very different environmental performances, namely recycling and recovery, are combined into a single target figure; and
- no account is taken of the prioritisation and performance against the Materials Hierarchy; in particular, it completely ignores Household Waste reduction, the most important aspect of the Materials Hierarchy.

An alternative way of measuring success

The Materials Hierarchy (the Three Rs) states, in order of decreasing benefit:

- reduce (including reuse and repair);
- recycle; and
- recover (both organics and energy) with organics recovery having a higher environmental benefit than energy recovery through EfW incineration.

It is important that any future targets reflect the relative importance of the Three Rs, if these targets are to be motivators for change. Unfortunately, we cannot measure the first and most important priority, that of Reduce, because we cannot measure the impacts of reduced consumption, or increased reuse or repair.

But, we could, for example, have one target which states that Household Waste arisings are not to exceed, say, 2017 levels. As the UK population increases, Household Waste reduction measures would thus have to be having an effect in order to achieve this target.

We could then have a second target which deals with the methods of treatment of the Household Waste that does arise, a target that emphasises the importance of recycling and organics recovery over EfW incineration.

This would give us three targets for Household Waste management, for example:

- the level of arisings to be no higher than that of 2017 (i.e. 27.5 million tonnes);
- the recycling rate to be at least 40%; and
- the organics recovery rate to be at least 30%.

This would revise the "45% by 2020" target to 70%, with the overriding cap stated in the first target.

However, I think the original UK government target of "45% by 2020" was flawed and, in particular, fell into the trap of being a target that reported success in increasing certain methods of treatment, namely recycling and organics recovery, but ignored others, the primary one being Reduction and to a lesser extent EfW incineration. I believe this sent the wrong signals, focusing people's attention, in particular, on driving up (albeit modestly) the recycling rate.

So, I'd like us to repurpose an approach that the homelessness charity Crisis has adopted, which focuses on what we are trying to drive **down**, rather than that which we are trying to drive **up**. Crisis is seeking to end homelessness, but rather than reporting the number of people successfully moved out of homelessness, their key target is to report on the number of people who **remain** homeless, with the intention of driving **down** this figure. They define their goal as "Functional Zero", the point at which homelessness has been effectively ended (zero will never actually be achieved, as people will always become homeless in the future and will need to be helped). Their approach is called Built for Zero and sets out their approach to drive homeless numbers down towards Functional Zero. The closer they get to Functional Zero, the more the programme has succeeded.

I would like to see the UK adopt a similar approach to implementing the Three Rs.

This brings me back to the issue of EfW energy recovery. If our target is to hit the 90% success rate shown in Figure 11.3, this will leave nearly 5.5 million tonnes (or 25%) of Household Waste being sent to EfW incinerators and nearly 1.9 million tonnes (or 5%) to landfill, a combined total of 30%.

Given that EfW incineration is only marginally better than landfill in environmental terms, I think this combined figure of 30% could give us our equivalent of the Crisis Functional Zero. We should therefore have a fourth target that states we will drive down:

- the treatment of Household Waste by EfW incineration or landfill to be no more than 30% of Household Waste arisings.

Conclusions on targets

We should have four Household Waste targets, not one:

- the level of Household Waste arisings to be no higher than that of 2017, i.e. 27.5 million tonnes;
- the recycling rate to be at least 40%;
- the organics recovery rate to be at least 30%; and

- the tonnage being sent to EfW incineration and landfill to be no more than 30% of total arisings.

The art of the possible

Could we achieve the above targets? Yes, I believe we could, and one of the things that gives me confidence is the recent example of what was achieved during the 2020 Coronavirus pandemic response, which showed that the previously unimaginable can be done.

Whilst our "waste" crisis does not have the immediacy of the pandemic and it is not a direct threat to health and life itself, it is a very serious issue that needs to be tackled now.

To show what is possible, who would have thought that during the Coronavirus pandemic:

- eight Nightingale hospitals could be created with a total bed capacity of 13,500 beds in less than two weeks each;
- teams of companies from very different backgrounds (from Formula 1 racing to car and aircraft manufacturers) could come together with academics and NHS experts to design and manufacture new respiratory ventilators in a matter of weeks;
- 90% of the rough sleepers in England could be housed, not in hostels, but in individual accommodation in just two days? (the "Everyone-in" campaign in 2020);
- companies that normally make fashion clothing could switch their production facilities to churning out Personal Protection Equipment (PPE) for hospital staff and care workers; and
- most unbelievably, research to develop a vaccine for the Covid-19 virus could be fast-tracked from years to months?

The previously unimaginable became reality because of the enormity of the threat facing the country. The drive for this unprecedented change in thinking was led by the Government and we need the Government to recognise the huge issue we are facing now with Household Waste and the damage to our planet.

But as well as needing the Government to lead, we need all of the players involved, starting with producers and retailers, to come together to deliver a co-ordinated and collective change. We will need new infrastructure to be built to manage Household Waste in new ways and at a scale not previously seen and we will need attitudinal changes from all the players, including you and me, so that we start to treat our planet with the respect that it deserves.

Notes

1 Defra, "Official Statistics – UK statistics on waste", www.gov.uk, updated 11 May 2022, accessed 24 April 2023.
2 Defra, "Official Statistics – UK statistics on waste"
3 Defra, "UK Statistics on Waste", Table 1.1, 19 March 2020.
4 WRAP, prepared by Eunomia Research Consulting Ltd, "National Household Waste Composition 2017", Table 7, January 2020.
5 If WDAs choose to there is no reason why materials couldn't still be collected at HMRCs, if that is what local people wanted.
6 WRAP, prepared by Eunomia Research Consulting Ltd, "National Household Waste Composition 2017", Table 7, January 2020.
7 The only materials assumed to be sent to landfill are those that cannot be incinerated, because they simply won't burn.
8 I use the term "economically" to include environmental considerations. If the most economic transport option is chosen, that is siting MRFs, processing plants, EfW plants and landfill sites as close as possible to the sources of materials, then the environmental as well as the economic impact will be minimised.

12

A PROPOSED WAY FORWARD

If you've read all the way to get to here, well done, I've covered a lot of ground. Now I'd like to pull this all together and set out a proposed way forward for how we consume the Earth's resources in ways that allow us to continue to live the kinds of lives we have chosen, but which do not irrevocably wreck our precious planet.

In summary

I've summarised below the key elements of what I have been advocating:

- stop treating what we discard as "waste" and view it instead as valuable, but no longer wanted, materials;
- producers to design and manufacture products and packaging with-the-end-in-mind, in terms of reuse, repair and recycling;
- redefine the Waste Hierarchy as the Materials Hierarchy and implement it according to the Three Rs by:

  - Reducing Household Waste through:

    - consuming less by buying what we consume more carefully;
    - reusing packaging as many times as possible; and
    - repairing products to extend their lives;

DOI: 10.4324/9781003504757-13

- recycling the products and packaging that can't be reused or repaired to make available the materials contained in these products and packaging as raw materials for remanufacture (moving towards a more Circular Economy);
- maximise Household Waste recycling by implementing:

  - the kerbside collection of an expanded, **standard** list of co-mingled Dry Recyclable packaging materials collected in 240-litre wheeled bins;
  - the establishment of a network of local MRFs to accept and sort these co-mingled Dry Recyclables;
  - the building of additional reprocessing capacity to ensure all collected Dry Recyclables are reprocessed **within the UK** into replacements for virgin raw materials;
  - the phasing out of bring collection points for Dry Recyclable materials that are to be collected at the kerbside;
  - new collection and dismantling arrangements for unwanted products, through retailer, WCA and WDA take-back and the creation of a new dismantling industry;

- recovering organic materials through kerbside collection, followed by:

  - anaerobic digestion; or
  - composting;

- maximising the recovery of organic materials by:

  - separate, weekly kerbside collection of food "waste" from ALL households;
  - fortnightly kerbside collection of garden "waste" (paid for by those householders who want such a service) by ALL WCAs;

- as a last resort, recovering unwanted, combustible materials that cannot be easily recycled or organically recovered using EfW incineration; and
- accepting that some unwanted materials cannot be recycled or recovered and so landfilling them.

National targets

We need to define what we want to achieve, by setting a number of targets. To date, the UK Government has been bound by EU targets on recycling and organic recovery, for example the recycling and recovery target of 45% by 2020. Now the UK is no longer a part of the European Union, it is free to set its own targets. I have already suggested the targets summarised in Table 12.1 on p. 215.

TABLE 12.1 Proposed UK targets

Method of treatment	Target
Reduction: • reduced consumption • reuse • repair	The annual level of UK Household Waste arisings to be no higher than in 2017, that is, 27.5 million tonnes
Recycling	The annual UK Recycling Rate to be no less than 40% of total Household Waste arisings
Recovery: • organic material • EfW incineration	The annual UK Recovery Rate for organic materials to be no less than 30% of total Household Waste arisings The annual UK Recovery Rate for EfW incineration to be **not more than** 25% of total Household Waste arisings
Landfill	The annual UK rate for landfilling of Household Waste to be **not more than** 5% of total Household Waste arisings

A revised national approach

As I said in Chapter 10, "waste" management is a responsibility that is devolved to the four governments of England, Scotland, Wales and Northern Ireland; there is no overall UK "waste" management strategy. But for the reasons I have already mentioned, we do need a consistent approach to how products and packaging are designed and manufactured in (or imported into) the UK and then how they are collected and treated.

I would love to see a UK Household Waste management strategy that sets out a clear vision and a practical solution for how we can sustainably consume and then manage the materials inherent to our consumption. Cannot the four devolved Governments come to a consensus?

We need a UK-wide approach because there are too many players involved, all with their own views and some with vested interests, for change to just happen. The required changes need to be driven and to do this we need to have a clear national strategy to deliver the stated targets, so everyone knows what we're trying to achieve and why.

Then we need a clear plan for how we are going to make it happen. But to bring about a societal change like this also requires leadership. We're talking about every person in the UK needing to think and act slightly differently; every local authority to review and potentially change how they operate with regard to Household Waste; we need producers and retailers to change how products are made and packaged and how unwanted products are collected and treated; we need the "waste" management and material reprocessing industries to invest in new sorting and treatment capacity and we need our Governments to legislate or regulate to make certain things mandatory. So, who is going to take the lead?

In the past, the European Commission provided a form of leadership by passing EU Directives that had to be enshrined in UK law. Now we no longer have the European Commission to drive this type of change, we have to look to the UK and devolved Governments to implement the legislation required to make such change happen. Without legislation, it will be impossible to corral all of the players and vested interests into implementing a comprehensive and cohesive plan of action. And action we need; not debate, not discussion, not prevarication and delay, but well thought-through and decisive action.

But Governments don't like to legislate for these kinds of change, they prefer industry to self-regulate and to find the solutions themselves. The closest we have come to legislating for changes to Household Waste are the introduction of the Landfill Tax in 1996, the introduction of the 5p plastic carrier bag levy in 2021[1] and the change announced by the then Chancellor Rishi Sunak stating that from April 2022 all plastic packaging was to be made with a 30% recycled content.

But legislation is only the start. The legislation has to be implemented and because there are so many players involved, all with vested interests, perhaps we need a new national authority with the powers to implement and police these changes. Whilst such an organisation could perhaps be funded by producers and retailers and/or from the proceeds of the Landfill Tax/Incineration Tax, it would need to be independent and it would need to have powers derived from legislation to make things happen.

But, whilst we need the UK and devolved Governments to take the lead, you and I can also make a difference and an immediate difference. We as consumers need to both change our ways of consumption, but in addition we can lobby producers, retailers and local authorities to change theirs.

We simply have to stop buying products where the materials from which a product or packaging is made are unacceptable. And we have to bombard producers and retailers with the message (email is powerful) that we won't buy Product X or Product Y, because it or its packaging are made from non-reuseable, non-repairable or non-recyclable materials. Consumer power is the most influential force to change the actions of producers, retailers and local authorities. Yes, legislation would really help, but don't hold your breath, do your bit now.

Some changes will take time to implement, but it is important to recognise that some changes could and should start today or within the next year. We can't wait five years or ten years, just look at our oceans. And you and I can change our buying patterns from tomorrow. I despair when I see targets being set for change in 2030 or beyond. That is far too slow. Change needs to start now.

A national strategy and plan

As I've already said, to implement a national strategy and plan will require a number of players to change how they currently operate and behave.

The players

There are ten players, yes ten, who need to do their part if we are to change consumer behaviour and move from a linear to a more circular approach to materials management. They are:

- durable product manufacturers;
- consumable product manufacturers;
- packaging manufacturers;
- retailers;
- the consumer or householder, that is you and me, so basically everyone;
- our local WCA (our local District, Borough or Unitary Council);
- our local WDA (our local County or Unitary Council);
- the waste management industry;
- the material reprocessing industries; and
- the Government.

Quite a list of people and organisations who all have a role to play. I've summarised below, what each of them will need to do, starting with the UK and devolved Governments.

Government responsibilities

I would like to see the UK Governments:

- require all WCAs to adopt a common approach to the collection of Household Waste, which is, co-mingled collection and MRF sorting of a standard list of Dry Recyclable materials;
- adopt a standard definition of recyclable materials to mean only those materials that:
  - are collected at scale:
    - for products this means take-back schemes must exist for the unwanted product via retailers (see Chapter 6); and
    - for packaging, the materials must be collected by all Waste Collection Authorities in their kerbside recycling collection schemes;

- can be separated into individual materials:

  - for products, a dismantling infrastructure must exist for the particular product; and
  - for packaging, this must be capable of being sorted into individual materials in a MRF; and

- are reprocessed to become the equivalent of virgin raw materials, economically and at scale.

- require all WCAs to provide effective food "waste" collection and treatment to all households in the UK (this is in process);
- ban the export of any part of Household Waste, so that **all Household Waste is processed within the UK**, including the re-processing of separated materials from MRFs; the UK should clean up its own mess and not export it to other countries that may not have the same high environmental standards or controls as the UK;
- address the issue of excess EfW incineration capacity in the UK by:

  - placing a moratorium on any new or the expansion of existing EfW plants; and
  - implementing an Incineration Tax, similar to the Landfill Tax, to make it more expensive to burn Household Waste, so as to overcome the arguments that reuse, repair and recycling are too expensive compared with incineration;

- discourage the use of non-recyclable materials in durable product and packaging manufacture by legislation or incentives to producers, in particular to end the use of:

  - minority plastics (minority by volume in Household Waste) that are hard to sort and reprocess, e.g. Polypropylene (PP) and Polyvinylchloride (PVC); and
  - plastics that are financially and and/or environmentally unattractive to recycle, such as Expanded Polystyrene (EPS)[2] and composite materials such as drinks cartons and other composite packaging.

Producer responsibilities

We need producers to design, manufacture and package goods differently. I've explained why and how they can achieve this, but there are three reasons they will be reluctant to change.

First, reuseable packaging and repairable products are likely to be more expensive to produce and producers are cost-driven, so they will resist any changes that increase their costs.

Second, making products from a more limited range of the materials that can be reused, repaired and recycled may well be more difficult for producers than simply using whatever material is the most convenient for manufacture. But it is no longer acceptable to just optimise the manufacturing process, producers have to **design-with-the-end-in-mind**, they have to think about what will happen to their products when they are no longer wanted and take responsibility for facilitating their end-of-life treatment.

Thirdly, if we as consumers buy fewer products, some producers will struggle to make the change from "pile-it-high-and-sell-it-cheap" to "improve-the-quality-and-sell-it-for-more"; I would like to see producers return to the days of taking real pride in everything they produce and to produce products of quality that last. But that's just me; maybe I'm naïve.

Thus, to reduce Household Waste, we need producers to design products that can be reused, repaired and recycled, even if this makes the product initially more expensive. We also need them to implement the infrastructure required for packaging reuse and product repair, as well as that needed to dismantle complex products to facilitate single material recycling.

We also need producers to use sustainable materials in their products, rather than materials such as plastics or, worse, composites of different materials, which cannot be easily recycled. This applies to both the products themselves and their packaging.

And we need producers to design products for recycling either by limiting materials to one material per product or by making products so that different materials can be easily separated for recycling (this applies to both products and packaging). Let's get product and packaging designers thinking in new ways, designing-with-the-end-in-mind as well as for the life of the product.

We also need producers to clearly and honestly label materials so it is obvious how they should be treated once they are no longer wanted (see Chapter 8 – Packaging labelling).

Retailer responsibilities

Bricks and mortar retailers

We obviously need all retailers to stock and promote products that are reuseable, repairable and recyclable. But we also need them to:

- encourage consumers to reuse packaging by selling products that are suitable loose, for example vegetables and dry foods and allowing the customer to bring their own reuseable packaging to take the products home; this obviously only applies to bricks and mortar retailers rather than online retailers, but it's also not just food, this could also apply, for example, to DIY hardware such as nails and screws;

- stop running "reuse" trials and really implement reuse approaches, i.e. stop greenwashing;
- take back unwanted products as part of selling a consumer a new, replacement product and transfer these unwanted "old" products to the original producer's designated repair/rejuvenate/recycle agent;
- stop wrapping products in multipacks that add an extra layer of packaging, or if this is done, only use recyclable packaging, for example, when two boxes of tissues are sold together, wrap them in paper, not plastic film; and
- revisit the criteria for use-by dates to reduce the amount of perfectly good food that becomes food "waste", simply because of our obsession with absolute dates.

Online retailers

There are three simple things that online retailers could do:

- use cardboard cartons and paper envelopes to package shipped products, instead of plastic packaging; I'm fed up with the number of deliveries I receive in Low Density Polyethylene (LDPE) envelopes, glibly printed with the statement "Fully recyclable" or "Recycle at Larger Stores";
- use gummed paper parcel tape to seal cardboard boxes instead of plastic adhesive tape, that is non-recyclable[3] (and not to use gummed paper parcel tape that is reinforced with plastic threads that increase its strength); and
- to provide shock absorption, use scrunched up paper packaging within the cardboard box, instead of plastic bubble wrap or plastic "quilt" (air-filled plastic bags), both of which are made from Low Density Polyethylene (LDPE) film, which is currently non-recyclable, because WCAs won't accept it in kerbside collection schemes.

But we should also be aware that the increasing move by consumers to the convenience of online shopping is driving a huge increase in the use of single life packaging. Just think how many cardboard boxes and plastic carrier bags are being consumed by home deliveries.

Consumer responsibilities

As consumers and householders, you and I need to:

- consume less and buy better (and buy-with-the-end-in-mind);
- only buy products (and their associated packaging) that are reuseable, repairable or recyclable; and
- actually reuse, repair and recycle products and packaging when we no longer want them (including separating unwanted materials responsibly for recycling).

To help you with this, I've included a detailed list of the things you can do as a consumer and householder in Annex II.

Local Authority responsibilities

Waste Collection Authorities (WCAs)

Remember, your WCA is your local District, Borough or Unitary Council.

Our Household Waste is collected from our homes, usually by a private contractor who works for our local council. But it is the local council that decides the collection arrangements and, in particular, the recycling arrangements that we as householders are then asked to use.

As I have already said, I propose that all WCAs implement a standardised, simple, easy to use and understand Household Waste recycling scheme comprising:

- a large, wheeled bin for a standard list of mixed recyclables;
- a smaller wheeled bin for non-recyclable "waste" for disposal;
- a food "waste" collection caddy; and
- a separate garden "waste" wheeled bin for those households that want one.

WCAs need to explain very clearly what should and what should not be included in the Dry Recyclables bin. Having a standard, national system will greatly help householders' understanding of what is required. This is both to maximise the amount of materials collected, but also to avoid the situation where some householders put as many items as they can into the recycling collection containers, hoping they will be recycled, even if the WCA hasn't asked for them (so-called "wishcycling"). Unwanted items won't be recycled and can cause collected loads of Dry Recyclable materials to be rejected on the grounds of contamination.

Waste Disposal Authorities

Your WDA is your local County or Unitary Council who lets the contract(s) for recycling and "waste" disposal. These contracts need to support my proposed Materials Hierarchy and so need to specify that the successful contractor must provide one or more MRFs to receive co-mingled recyclables from the participating WCAs and that these MRFs have to be capable of effectively separating the standard list of co-mingled materials and the MRF operators have to have contracts in place for the reprocessing of the separated materials within the UK.

Given the overcapacity in the EfW sector, we need WDAs to work together to implement co-ordinated EfW disposal deliveries that allow the guaranteed tonnages within existing EfW contracts to be honoured, whilst allowing Household Waste to be reused, repaired, recycled and recovered. And we need WDAs to not let future EfW contracts that tie them into minimum tonnages that will limit the amount of Household Waste being treated by reuse, repair, recycling or organic recovery.

Waste Management Industry responsibilities

There are three areas where we need the waste management industry to provide sufficient, local infrastructure to support the collection activities of WCAs and for WDAs to contract for such facilities as necessary. These three areas are:

- to support the co-mingled collection of Dry Recyclables, we need local MRFs to be built to sort and sell on separated materials to material reprocessors (this will include the design of new MRF technology to separate the designated Dry Recyclable materials to include materials such as plastic film, which is not currently collected and sorted);[4]
- the building and operation of local anaerobic digestion facilities for food "waste" treatment in the right places to service the needs of local communities; and
- the building and operation of local composting facilities for garden "waste" treatment that provide a local outlet for WCA collected garden "waste".

Reprocessing Industry responsibilities

We will only achieve the maximum level of recycling that is possible, if we have sufficient reprocessing capacity available to reprocess all the collected materials. Plus, there needs to be sufficient reprocessing capacity for each material within an economically viable distance of the MRFs that will generate the materials for reprocessing. What we are talking about are paper mills, glass reprocessors, aluminium and steel reprocessors and plastics reprocessing plants. Whilst paper mills exist within the UK, some "waste" paper is exported for reprocessing; this should stop as it is environmentally damaging to do this. We need to have sufficient reprocessing capacity within the UK to reprocess all of our Dry Recyclables.

Education

But as well as each of the players involved doing their part, if we are to bring about societal change, we need to educate consumers in order to influence the way they behave. We have to show them why we need to change, how we can change and what the benefits would be.

And in terms of consumers I suggest that education needs to be at two levels: children and adults, but as well as being consumers and householders, some adults are involved in the Supply Chain and so can take their new found understanding into their workplace.

Educating children

It is vital we teach our children about the dangers facing our planet and how we need to behave differently to mitigate these dangers. We must show them what needs to be done and why. Many children already understand this and are gravely concerned about what we are doing to our planet (just think of the response among young people to Greta Thunberg's leadership on climate change).

I think we must educate them for three reasons:

- they will become the consumers of the future and we need their behaviours to be the right ones;
- when they grow up they will become the managers of the Supply Chains of the products we consume and of the processes that manage what happens to products and packaging once we no longer want them, so we need them to be thinking and acting in the right ways; and
- they are often the best teachers of their parents and can influence their parents' behaviours, and I'm thinking here particularly of their parents as consumers and householders and but also as today's managers of the Supply Chain and the processes that manage what happens to no-longer-wanted products and packaging.

Children really care about our planet and what is happening to it, so it is incumbent on us to teach them what is happening and how we can change to improve things. They will listen and even if they can't effect change themselves, they can be very vocal and persistent advocates of the right behaviours.

Educating adults

We clearly need to educate adult consumers about the need to and how to consume differently. But we also need to educate and retrain today's managers of the Supply Chains of the products we consume and the processes that manage what happens to products and packaging once we no longer want them. We need product designers to design in new ways, retailers to promote products and packaging to support the new ways of consuming and we need producers and retailers to manage our no-longer-wanted products in new ways.

For the first of these groups, product and packaging designers, I would dearly like to see product and packaging design education and training courses at colleges and universities teaching young people who are coming into the design profession to think differently, to move us as close to Circular Supply Chains as we can get. We need them to understand the importance of designing-with-the-end-in-mind, of the roles reuse, repair and recycling should play and of the Circular Supply Chain itself.

Final conclusions

If you take away nothing else, these are my key points

If I have to boil down everything I've said in this book to just four things they are:

- producers need to design-with-the-end-in-mind and make products and packaging that are reuseable, repairable and truly recyclable;
- producers and retailers must implement a simple take-back infrastructure for the repair and recycling of products;
- local authorities must give householders a simple means to recycle their no longer wanted materials which means implementing co-mingled, kerbside collection and MRF sorting of packaging and the separate collection of food and garden "waste"; and
- the UK Governments should legislate to ban the export of Household Waste collected for recycling, place a moratorium on the construction of new or the expansion of existing EfW incinerators and impose an Incineration Tax at least for Household Waste.

Can we change?

To make our consumption of the Earth's resources sustainable, many players in the Supply Chain have to change to create Circular Supply Chains.

It's a big ask, so can it be done? Yes, I think it can. There is a developing awareness of the need for change and a growing appetite for change, but we need four things to make this happen:

- a real acceptance by everyone that we cannot go on trashing our planet;
- a clear blueprint for a new approach to consumerism, backed up where needed, by clear and targeted legislation and regulation to make some changes mandatory;

- a real readiness from all players in the Supply Chain to change behaviours and actions; and
- a spirit of co-operation and working together by all players in the Supply Chain to bring about the changes we need.

The first of these is slowly starting to happen. People are beginning to wake up to the horrors we are inflicting on our planet, for example the public reaction to the appalling revelations of plastic pollution in our oceans as shown on the television series *Blue Planet II*. But this is slow and, if I'm brutally honest, many people just don't care.

I hope this book is a solid start to the second point and that our UK Governments and all the other players will take this to heart.

The third is very slowly, too slowly, starting to happen, but there is a reluctance on the part of some players, notably producers and retailers, to change, in large part driven by cost concerns. This is one of the reasons why we as consumers need to make our voices heard.

And on the fourth point, let's draw hope from the responses to the 2020 Coronavirus pandemic. This was an unprecedented challenge to all parts of our society and we saw previously unimaginable responses from many areas. When the chips are down, unimaginable change and co-operation can happen.

Well, the chips are certainly down for our planet. Okay we aren't facing the immediate risks of illness and death from Coronavirus, but we are putting our future survival at risk by the way we are treating the Earth. And the need to act is urgent. We don't want empty rhetoric and we don't want plans that will take five or ten years to enact; we need action from today and real change happening within two years.

I support the comments made by our late Queen when she was caught on microphone in October 2021, privately saying "… it is very irritating when they talk, but they don't do" whilst she was discussing attempts to tackle Climate Change.[5] And again, her grandson Prince William told the BBC on the same day that the world's greatest minds need to be "… fixed on trying to repair this planet, not trying to find the next place to go and live [meaning another planet]". He also apparently said that there could not be "… clever words but not enough action" at Cop26.[6]

As I'm sure you know, Prince William launched the "Earthshot Awards" in 2021; the title of the awards apparently references President Kennedy's announcement in 1961 of a "Moonshot" when he said that "I believe that this nation should commit itself to achieving the goal, before this decade is out, of landing a man on the moon and returning him safely to Earth". America achieved this goal in 1969 and today we take space flight and the landings on the moon for granted. So much can be achieved when we put our minds to it.

I've set out a comprehensive plan for how we can achieve very significant improvements to how we treat unwanted materials, once we as consumers have finished with them. If we can achieve something like this plan and do it in less than, say, five years, then I think this would be progress worthy of an Earthshot award. I think we need to see far more radical proposals and action, otherwise we'll be guilty (as Prince William said) of "… clever words but not enough action".

Tick tock

The clock is ticking and it is no exaggeration to say the Earth is in grave danger. As David Attenborough said in *A Life on Our Planet*, the Earth has seen five catastrophic upheavals that we know of, all of which caused widespread destruction and mass extinction.[7] The last such upheaval saw the extinction of the dinosaurs. Today we are witnessing the sixth upheaval, named the Anthropocene, but unlike the previous five, this one is entirely man-made. All is not lost, yet, as there is still time to stop this becoming the sixth cause of planetary destruction and mass extinction, just.

Notes

1   "Carrier bags: why there's a charge", www.gov.uk, 21 May 2021, accessed 25 April 2023.
2   Two of the key attractions to producers of EPS are that it is lightweight and very difficult to crush or compress, but these characteristics make it very costly to transport and reprocess. Instead, producers can use moulded cardboard packaging (some already do), made in the same way that cardboard egg boxes are.
3   If Amazon can, everyone can.
4   There needs to be a sufficient number of conveniently located MRFs to allow collected co-mingled recyclables to be separated as close to the point of collection as possible, but also within an economic distance of the various reprocessing facilities required.
5   "Queen's 'talk, no action' criticism over Cop26 caught on microphone", *The Times* newspaper, 15 October 2021.
6   "Queen's 'talk, no action' criticism over Cop26".
7   "What is the sixth mass extinction and what can we do about it?", www.worldwildlifefund.org.com, 15 March 2022.

ANNEX I

How individual materials are reprocessed

Organics

Food "waste"

The WRAP website describes the anaerobic digestion process as follows:[1]

Anaerobic digestion processing systems operate in different ways. For example, material may be fed into a reactor in distinct batches, or in a continuous flow.

In this guide we detail the basic steps in the AD process. Although the exact process may vary for different types of AD or be altered to favour intended outcomes, most current AD operations follow this process.

The process

Pre-treatment:

- Feedstock processing begins with pre-treatment. This involves mixing the various feedstock elements together to ensure the right consistency and C[arbon]:N[itrogen] ratio and may also involve the addition of water.
- The material should also be screened for contaminants, such as plastic[2] and grits at this stage.
- Packaged food waste will also be extracted from its packaging at this stage.

Digestion:

- The feedstock is then fed into a digestor.
- Digestion usually occurs under a slow mixing process and the biogas produced is collected.

Use of the digestate:

- The digestate produced is stored until required, and can be separated into liquid and solid fractions.
- Solid fractions can be processed further on site by being put into a composting operation for further processing or used directly on land.
- The liquid may also be used on the land as a biofertiliser.

Use of biogas:
The biogas produced will be stored before being either refined further into biomethane for vehicle fuel or for injection into the gas grid or burned in a combined heat and power engine to produce electricity and heat, or burned in a gas boiler to produce heat for local use.

Garden "waste"

Commercial scale composting is usually carried out as follows:

- incoming green waste (garden waste such as prunings, clippings, weeds, leaves and discarded plants) is hand checked for contaminants such as plastic bags[3] and large items such as tree stumps are removed (food waste is excluded from green "waste" collections as it can attract vermin to the composting site);
- the green waste is then shredded to reduce everything to a reasonably uniform size and the material is laid out in long rows called "windrows" on a concrete base, with each windrow typically containing 400–500 tonnes of material.[4] The material is composted for six to twelve weeks;[5]
- the material is turned regularly (up to three times per week) using specialist equipment during the time it takes to complete the composting process.[6] Temperatures are monitored daily to ensure the material remains above 60°C (140°F) to kill off weed seeds and pathogens; and
- finally, the matured compost is screened down to a uniform size and bagged or sold loose and is sometimes used as landfill cover.[7]

Dry recyclables

Paper and cardboard

Collected paper and cardboard are baled at the point of collection (either the point where paper banks are emptied or at a MRF) and then shipped to a paper or board mill for processing.

Here the material undergoes a four stage process:

- screening and sorting: the incoming material is manually checked for major contaminants and may be sorted into different paper grades, depending on the end product being produced;
- shredding and pulping: after shredding to reduce the size of the paper pieces, large quantities of water are added, together with chemicals that help the process of breaking down the paper, a process known as pulping; after pulping, the material passes through a series of screens and centrifuges to further remove contaminants such as metal staples, adhesive tapes and plastic film;
- de-inking: the pulp then passes into floatation tanks to remove dyes and inks (which have to be disposed of); at this stage the pulp is 99% water and 1% paper fibres; and
- the pulp is then transferred to the paper-making machine, which comprises a series of heated rollers that produce rolls of finished paper up to 30 feet wide and weighing up to 30 tonnes.

The one exception to the above process is the manufacture of moulded paper/cardboard products such as the ubiquitous egg box or mouldings to protect delicate products, instead of moulded plastic. The process is similar to the above, except instead of the fourth step of the paper machine, the pulp is passed into moulds to form the required end product.

Glass

Collected glass containers (bottles and jars) can be reprocessed into new containers again and again, with no loss in quality and with significant energy savings (about a 40% saving). But to produce high quality new containers, the glass must be separated by colour, i.e. into clear, green and brown.

As well as colour sorting, it is essential that contaminants, such as ceramics, porcelain/china and non-container glass (such as drinking glasses or window glass) are removed. Contaminants can lead to inclusions or holes in the new glass containers.

When collected, whether via kerbside or bottle banks, the container glass colours are mixed (although a minority of bottle banks are colour separated).

After visual inspection to remove obvious contaminants, the glass containers are broken up. The broken glass then passes under magnets to remove steel lids and bottle tops and eddy current separators to remove non-ferrous bottle tops and enclosures. People often ask whether bottle tops and jar lids should be left on bottles that are to be recycled. From the foregoing, the answer is obviously yes (except corks in wine bottles which should be removed).

The broken glass is colour separated by scanners which are very accurate. The resulting glass cullet is then mixed with raw materials comprising silica sand, sodium carbonate and limestone before being melted in a furnace at about 1,550°C. The resulting molten glass is then manufactured into new glass bottles and jars.

But as well as collected glass bottles and jars being recycled into new bottles and jars, some glass, e.g. that collected from incinerator ash, is processed to form aggregate for use in construction works, such as being used in the base layer for roads. But this is not recycling, it is not circular, rather it is recovery.

Metals

Aluminium[8]

Collected aluminium cans are baled at the point of collection or separation (for example at a MRF) for ease and efficiency of transport. Upon arrival at the Novalis plant in Warrington the bales are broken open and the cans shredded into pieces the size of a 50p piece. These pieces of aluminium are then passed through an oven that burns off any printing ink, paint or lacquer (from the inside of the can), before being loaded into a furnace operated at 650°C–750°C for melting. The liquid aluminium is cast into 27-tonne ingots, which are then hot and cold rolled into sheet for reuse as cans, foil or other aluminium products.

According to Novalis approximately 10 billion aluminium drinks cans are sold in the UK each year. Each can weighs 17g,[9] so that's 170,000 tonnes of recyclable aluminium cans per annum.

Steel

Steel cans and other ferrous (steel) products that are collected for recycling are reprocessed via the existing and extensive scrap metal infrastructure that exists in the UK. A strong market exists for ferrous scrap. Steel products are separated from non-ferrous products, predominantly using the fact that steel can be magnetically separated using electromagnets.

The ferrous scrap is then graded and added to virgin material for smelting.

Plastic

Plastic products and packaging comprise two types of plastic: rigid plastic (bottles, trays and, for example, electrical or electronic product casings) and flexible plastic (plastic film).

According to the industry body RECOUP, all WCAs collect plastic bottles and 87% collect pots, tubs and trays (rigid plastic). In 2021, 61% of plastic bottles and 36% of pots, tubs and trays were apparently collected for recycling.[10] Interestingly, the figure for plastic film collection for recycling was just 4%, given that almost no WCAs include plastic film in their kerbside collections for recycling.

Collected mixed plastics are taken to a MRF for separation from other materials and then to a Plastics Recycling Facility (PRF). Here the plastics are washed and shredded prior to separation into polymers using optical sensors. The sorted polymers are then melted and extruded to form pellets that can be used to manufacture new plastic products.

Alternatively, mixed plastics with a high proportion of polyethylene are shredded, melted and extruded into mouldings to be used in the manufacture of products such as fence posts or simple furniture.

Textiles

Whenever we talk about textiles, we tend to think of clothing. But textiles are used much more widely, in our homes, in our places of work and leisure venues, in vehicles and in many places besides.

As well as consuming virgin raw materials, textile production also consumes vast amounts of water[11] and energy and is a major contributor to climate change and pollution.[12]

Textiles are usually collected via textile or clothing banks, rather than by kerbside collection, with the UK collection rate being estimated to be 37%.[13] Collected textiles are baled prior to shipment to reprocessing facilities.

Collected textiles are reprocessed as follows:[14]

- manual sorting and grading into:

  - wearable textiles (including shoes), for reuse within the UK or exported, often to third world countries;
  - unwearable textiles, for shredding and re-spinning into yarn;

- unwearable textiles are graded by material type and colour, prior to:

  - shredding and pulling into fibres;
  - carding to clean and mix the fibres; and
  - spinning ready for subsequent weaving or knitting.

WEEE

The amount of WEEE that we generate

According to a report[15] from PACE (the Platform for Accelerating the Circular Economy) cited at the World Economic Forum in Davos 2019, the world produced 50 million tonnes of WEEE in 2017, a figure equivalent to all the jumbo jets **ever** made and a figure that could rise to 120 million tonnes by 2050. The report states that only 20% of this WEEE is safely recycled, with the remaining 80% either being landfilled or dangerously treated in developing countries like Nigeria where workers are exposed to the toxic chemicals contained within WEEE. It's a massive issue, exacerbated by the fact that many of the materials contained in WEEE are extremely valuable due to their scarcity.

The products that make up WEEE

Waste Electrical and Electronic Equipment comprises a wide range of unwanted equipment, from fridges and freezers to computers and mobile phones. The materials they contain are predominantly plastics, glass and metals. Common metals such as copper, lead, steel and aluminium are present, as are precious metals such as gold and rare earth metals. It appears that whilst there is great interest in recycling these materials, in particular the rare earth metals, which are in huge demand, the WEEE recycling industry is relatively new and has yet to become really established.

In addition, WEEE contains a number of toxic elements, not least lithium-ion batteries, so that the recycling operations need to be carried out under very controlled conditions.

The processing of WEEE uses a combination of manual sorting, mechanical separation (trommel screening, magnetic collection of ferrous metals, eddy current separation of non-ferrous metals) and chemical metal extraction.[16]

But, it is undoubtedly the case that this industry will develop, not least because the concentrations of rare earth metals in WEEE is much higher than in naturally occurring ores.[17]

Untreated wood

Untreated wood is wood that has not had paint, varnish or any other finish applied to it. It is in effect just wood. Untreated wood is crushed and shredded for use either in wood products such as chipboard (but the wood must be clean for this process) or for inclusion in Refuse Derived Fuel (RDF) pellets, which are burned in specialist boilers.

Notes

1 "Anaerobic digestion: The process", www.wrap.org.uk/home/resources/guide, accessed 11 October 2022.
2 This is why plastic bags should not be used to line food "waste" containers as they will be rejected at the anaerobic digestion plant as contaminants and sent for EfW incineration or landfill, thus condemning this plastic film to energy recovery, when it could otherwise be recycled.
3 For more on this see Chapter 9 "Biodegradable plastics and why they are a bad idea".
4 "Commercial composting", www.biffa.co.uk, accessed 9 September 2022.
5 www.waste-technologies.co.uk, accessed 9 September 2022.
6 www.waste-technologies.co.uk
7 "Waste" that is tipped into a landfill site is covered, at least daily, with a layer of material which is described as "landfill cover". Landfill cover is required to minimise the opportunities for birds or animals to access the deposited "waste" and to minimise the risk of the lighter deposited materials being blown around the site.
8 www.novelisrecycling.co.uk
9 www.alucan.org.uk states 60 cans per kilo, accessed 12 January 2024.
10 "RECOUP 2021 UK Household Plastics Collection Survey", 2021.
11 "Textiles – important facts", https://bir.org/the-industry/textiles, accessed 30 September 2022. 10,000–20,000 litres of water are used to produce 1kg of cotton textile.
12 "Textiles – important facts", The World Bank estimates 20% of global water pollution is caused by textile processing.
13 European Clothing Action Plan, "Used Textile Collection in European Cities", March 2018, Table 2, p. 18.
14 "Textiles – recycling processes" https://bir.org/the-industry/textiles, accessed 30 September 2022.
15 "A New Circular Vision for Electronics – Time for a Global Reboot", www3.weforum.org/docs/WEF_A_New_Circular_Vision_for_Electronics.pdf, accessed 25 October 2023.
16 A. Marra, A. Cesaro and V. Belgiorno, "The recovery of metals from WEEE: state of the art and future perspectives", *Global NEST Journal*, 2018, vol. 20, no. 4, pp. 679–694.
17 A. Marra, A. Cesaro and V. Belgiorno, "The recovery of metals from WEEE: state of the art and future perspectives", *Global NEST Journal*, 2018, vol. 20, no. 4, pp. 679–694.

ANNEX II

Tips to help you to help the planet

What can I do?

You might think, but what can I do to make a difference? I'm only one in 69 million people in the UK and one in eight billion in the world. But as Greta Thunberg says "No one is too small to make a difference"[1] and look what she has achieved in raising the awareness of individuals and national governments. If you agree with what I am advocating, I ask you to live a more conscious life, lead by example and encourage others to change whether by just talking to family and friends or lobbying retailers, producers and the Government. If many of us say things have to change, but most importantly, how things have to change, then things will change. But if we remain silent, then nothing will happen.

We all need to personally follow the Materials Hierarchy which basically means doing four things:

- consume less;
- practice reuse;
- rediscover the repair of products; and
- set out our unwanted materials at the kerbside for recycling (Dry Recyclables) and recovery (food and garden "waste").

I have set out below a number of suggested things that you and I can do now that will make a difference.

Reduce

We all need to consume less. But when we do buy things, we should buy more mindfully and then break the throw-away habit. So:

- **Buy-less-stuff**. Only buy things that you really, really want or need. Stop being driven by fashion and buying for the sake of buying.
- **Borrow, don't always buy**. Do you know someone who has something you need for a short time, like an electric drill or a stepladder. If you do, then ask to borrow it rather than buying one yourself and consigning it to the back of the cupboard when you've used it. But here's a tip – take excellent care of borrowed items and return them quickly – do as you would be done by.
- With regard to packaging, **buy-with-the-end-in-mind**. Try hard to not buy products packaged in plastic and look for recyclable packaging such as paper, metal or glass instead. Other suggestions are:
  - avoid individually wrapped items bundled into a multipack using yet more packaging (nearly always plastic film); tins of baked beans are an example of this as are packets of biscuits and "handy-sized" packs of things like tissues; if tins of tomatoes can be sold in packs of four, wrapped in a cardboard sleeve, why can't other products?
  - buy less preprepared food, whether it's pre-chopped vegetables or full microwaveable meals; it all comes heavily packaged in plastic; try to at least occasionally cook from scratch using basic ingredients and limit the number of take-aways you eat, they too come heavily wrapped in plastic including the dreaded expanded polystyrene clam shells; but
  - please don't beat yourself up about this, try your best and if you can manage to buy better at least some of the time, that's a good start, there will be times when you can't avoid the overpackaged product or plastic-wrapped whatever-it-is-you-want, but at least try to find a more acceptable alternative if you can and let retailers know this is what you want.

- **Buy once, buy well and buy fewer, better things**. Don't just buy the cheapest product, buy good quality products that will last and, if appropriate, can be repaired and judge the cost of the item over its lifetime.
- **Repair, don't replace**. Higher quality products last longer and are potentially repairable, so costing less in the long run that the cheaper, disposable alternative. Use sites such as eBay to source spare parts to help you repair products.
- **Treat food "use-by" dates with scepticism**. Every year we throw away 6.6 million tonnes of food "waste" in our Household Waste. Some of this is food we don't want to eat, like potato peelings or fat trimmed from

meat, but a lot is perfectly good packaged food that is discarded because it has passed its "use-by" date. If you treat these dates as advisory, not mandatory, and use your eyes and nose to really test if the food has gone off, not only can you significantly reduce the amount of food "waste" you produce, but you can save yourself money as well. BUT, please be careful and don't take risks.

- **Buy-with-the-end-in-mind**. Think about what will happen to what you are buying when you no longer want it. Buy the products and packaging that will have the best outcome when they have reached the end of their lives. For example, buy foodstuffs packaged in paper, cardboard, glass or metal cans, rather than plastic, so the packaging can be recycled.
- There are also any number of small things you can do immediately that will make a difference and show good examples to others:
  - **Ban the bin-liner** (I know not everyone will agree with this). We extract crude oil, process it into Low Density Polyethylene (LDPE) film, manufacture bin liners from the LDPE film, fill them with our unwanted "waste" and bury them in a landfill or incinerate them. Bin liners are made to be disposed of. And every year we consume thousands of tonnes of these bin liners. But, if you have access to a kerbside food "waste" collection service, you don't actually need to use bin liners, because once food waste is taken out of your disposal wheeled bin, what is left is relatively clean and dry materials (mainly non-recyclable plastic film). These can either be collected directly into an unlined kitchen bin or into a kitchen bin containing a bin liner that is not thrown away when the kitchen bin is emptied into the wheeled bin for kerbside collection. And we absolutely do not need bin liners in wheeled bins. Take food waste out of the picture and what is left are relatively clean and dry materials.
  - **Cut-the-cling-film**. Stop wrapping everything in your fridge in single-use, non-recyclable, plastic film. If you need to cover food, try putting an upturned bowl or plate over it, wrap it in beeswax wrap,[2] or put it in a reuseable container such as a Tupperware box (and I use the word Tupperware to mean any type of plastic storage box, which might be a reused ice-cream container or plastic take-away box). I would love to see TV chefs encouraging viewers to **cut-the-cling-film**.
  - **Try gummed paper parcel tape**. This is one of my pet hates. I buy things on eBay and often the parcel comes completely wrapped in plastic sticky tape. This plastic tape is definitely non-recyclable and can only be got rid of by the householder as "waste" for disposal. If you are sending a parcel, please don't use plastic tape, replace it with gummed paper tape, which is just as effective. And this doesn't have to be gummed tape that has to be wetted before it will stick; self-adhesive

paper tape is available and is just as easy to apply (just look at your next Amazon parcel delivery!) But beware, some paper parcel tape has plastic "wires" embedded in it to make it stronger, so there is still plastic hidden within what appears to be a paper product.

- **Don't screw it up**. Some people screw up things like sweet wrappers or paper; please don't. The MRF sorting needs items to be as large as possible and screwing up things like sweet wrappers makes separating the different materials impossible.

- Finally, lobby for change:

  - If you see an example of bad packaging, please write or email the retailer and/or product manufacturer to request they change their packaging. A quick email to the Customer Services Department is all that is needed (the email address is usually printed on the packaging). I accept that if only I write it might not make a difference (but, in quite a few cases I have received a very positive response), but if many consumers write, I think it will make producers think hard about change. But if you do write, please suggest alternatives, don't just criticise.

  - Write to your local WCA and ask them about what happens to collected recyclables. Show that you are interested in what happens to your recyclables and hold them to account.

Reuse

- Reuse and refill:

  - Switch from buying milk in single-use plastic milk "jugs" from shops to having it delivered to your door by your local milkman, in reuseable glass milk bottles. Not only will this help with resource management, but it could create new jobs. And don't forget to recycle the aluminium tops as well.

  - Buy food loose if you can and take it away in your own reuseable containers, preferably not plastic containers or bags bought specifically for this purpose, as this is just consuming even more plastic. Instead, use natural fibre bags and reuse your existing storage boxes or repurpose used ice-cream containers, etc.

  - An obvious one: don't buy water bottled in single-use plastic bottles, instead refill a reuseable bottle from the tap (or even refill what appears to be a single-use bottle of water).

  - Review all the products you use and see if there is a reuseable/ refillable alternative. For example, if you use a fountain pen, why not switch from non-recyclable plastic cartridges to using ink from a glass bottle? Or switch from plastic liquid soap dispensers to a bar of soap that comes wrapped in a paper wrapper? Or buy butter in a wrapper and use a butter dish, instead of buying a (usually non-recyclable) plastic tub of butter? Switch to shampoo and conditioner bars instead of plastic shampoo bottles (you'll be surprised how good these are and how easy they are to find). The list goes on.

- **Look for refillables**. It's not just about milk and glass bottles. Some retailers are starting to offer refills to save packaging (not enough retailers and not enough products), so seek them out, but make sure the refill comes in a container that can itself be recycled (plastic pouches are bad news).
- **Don't-ditch-when-down-sizing**. If you declutter or downsize, don't just bin the unwanted things, sell them online or donate to them to sharing websites like Freecycle.com or charity shops or if none of these are possible then recycle them.

Repair

I've already said **Buy-fewer-better-things** and look for products that can be repaired. And when something wears or breaks either seek out someone who can repair it for you or have a go yourself; it's tremendously satisfying.

Recycle and recover

Recycle

- Make sure you understand what your WCA will and will not collect in your kerbside collection scheme and set out these materials and only these materials. Also remember what will be collected by the WCA when you **Buy-with-the-end-in-mind**.
- **Plastics or planet?** Try not to buy single-use plastic packaging, instead buy food and drink sold in recyclable containers such as cans, glass jars and cardboard boxes. For example, look for chocolate bars wrapped in aluminium foil, inside a paper outer wrapper, rather than a bar wrapped in plastic. Okay, your favourite chocolate bar may not come in foil and paper packaging, so email the manufacturer and tell them you are switching away from their product, because of the packaging.
- **Sweat the small stuff**. There is an American expression, "Don't sweat the small stuff" meaning, don't worry about little things. With recycling, I'm arguing the opposite. There are too many small items that can be recycled, we just have to be a little creative about how we collect them. So:
  - Please collect aluminium milk bottle tops. Each bottle top weighs approximately 0.2g[3] which sounds too small to bother with, but

hold on. Five billion litres of milk are sold for drinking in the UK each year[4] and about 3% of this is delivered to doorsteps,[5] so about 150 million litres or 264 million pints. Assuming this is all delivered in glass bottles, that is nearly 264 million aluminium bottle tops each year, equivalent to 53 tonnes of aluminium, equivalent to about three million aluminium drinks cans. An example of such milk bottle tops is shown in Figure A2.1.

- Separate the aluminium covering from pre-packed medicinal tablet blister packs. Someone who takes just two tablets twice a day generates 1,500 aluminium tablet tabs each year. Each one weighs only 0.02 grams,[6] but over a year, this adds up to 30g of aluminium foil per person. If we assume that everyone in England over the age of 60 (about 14 million people)[7] is taking four tablets a day (and many older people take far more than four pills a day, so the following is a very conservative estimate), this will send over 400 tonnes of aluminium for recycling **every year** in England alone, saving the mining of 1,600 tonnes of bauxite.[8] To put this in perspective, this would represent eight times the amount of aluminium foil that could be collected from milk bottle tops. Figure A2.2 illustrates what I mean by aluminium tablet tabs

- But you can't put such tiny pieces of aluminium in your kerbside bin and expect them to be recycled; they'd simply be lost. So, save them

FIGURE A2.1   Aluminium milk bottle tops. Photograph courtesy of Lily Waite.

FIGURE A2.2 Aluminium tablet tabs. Photograph courtesy of Lily Waite.

up and when you have enough, wrap milk bottle tops, tablet tabs, foil chocolate bar wrappers, aluminium lids from dairy pots such as yoghurts and even the foil from wrapped sweets, into large pieces of used aluminium foil or food trays. If you're not sure if the material is aluminium foil or plastic film, use the **fold-the-foil** test. If you fold and crease it and it stays folded, it's aluminium; if it springs back, it's plastic.

- The same applies to steel beer bottle tops. Stop them from getting lost in the MRF by putting a few inside an empty steel food can and then squashing the can with your foot to trap the lids inside. Figure A2.3 shows how this can be done.

- Basically **wrap-the-small-stuff**. Wrap small items in bigger pieces of the same material to stop the smaller items getting lost (e.g. put small pieces of paper inside discarded envelopes). But don't put till shop receipts in the recycling bin. These are printed on thermal paper which is not recyclable.

- **Separate different materials** from each other when you set them out for recycling collection. For example: some large yoghurt pots are made from a thin plastic pot, encased in a cardboard outer sleeve to give rigidity. These have a zip printed along a perforation showing you where to rip the cardboard to separate it from the plastic pot (why can't **all** large yoghurt pots be made like this?) But there is no need to take the steel lids off glass jars or the metal tops off wine bottles as these are separately collected when the bottles are crushed.[9] But do remove any non-recyclable materials from

FIGURE A2.3   Collect steel beer bottle tops. Photograph courtesy of Lily Waite.

those that are recyclable, for example take the non-recyclable plastic film off recyclable food trays and separate paper and cardboard from plastic packaging.

Recover

- If your WCA provides a food "waste" collection service:
  - please use this for all your unwanted foodstuffs;
  - separate unwanted food from its packaging, before placing these separately into your food "waste" caddy and your recycling container(s); and
  - only line your food "waste" cady with paper liners, not plastic.
- If your WCA doesn't provide a food "waste" collection service, ask them why not and keep asking them until they do (they are now obliged to do so).
- Compost your garden "waste" either:
  - by doing this yourself in a home composting bin; or
  - if your WCA provides a garden "waste" collection service, then taking advantage of this; and
  - if you have a lot of garden "waste" or material that is too woody to compost, then, if you can, take this to your local HWRC for collection.

Notes

1   Greta Thunberg, *No One Is Too Small to Make a Difference*, Penguin Books, 2019.
2   Beeswax wrap is essentially cotton cloth that has been coated in beeswax on both sides. When wrapped around foodstuffs or over the top of a container, the wrap sticks to itself, helping to keep it in place. After use the wrap can be lightly washed in warm soapy water, ready for reuse.
3   Author experiment confirmed by www.jacksonsdairy.co.uk, accessed 12 January 2024.
4   "Fascinating facts about British Dairy", www.countrysideonline.co.uk, accessed 20 January 2021.
5   "UK Dairy Industry Statistics Research Briefing", House of Commons Library, 1 May 2020.
6   Author experiment.
7   The population of England over 60 years of age taken from the www.statista.com website giving the results of the 2021 census was 13.7 million people, accessed 30 August 2023.
8   "UK Aluminium Industry Fact Sheet 17 – Primary Aluminium Production", www.alfed.org.uk, accessed 14 November 2023.
9   "You can leave your cap on!", www.alupro.org.uk, accessed 20 September 2022.

INDEX